D1801484

SCIENCE BEFORE SOCRATES

SCIENCE BEFORE SOCRATES

Parmenides, Anaxagoras, and the New Astronomy

Daniel W. Graham

OXFORD
UNIVERSITY PRESS

Oxford University Press is a department of the University of Oxford.
It furthers the University's objective of excellence in research,
scholarship, and education by publishing worldwide.

Oxford   New York
Auckland   Cape Town   Dar es Salaam   Hong Kong   Karachi
Kuala Lumpur   Madrid   Melbourne   Mexico City   Nairobi
New Delhi   Shanghai   Taipei   Toronto

With offices in
Argentina   Austria   Brazil   Chile   Czech Republic   France   Greece
Guatemala   Hungary   Italy   Japan   Poland   Portugal   Singapore
South Korea   Switzerland   Thailand   Turkey   Ukraine   Vietnam

Oxford is a registered trade mark of Oxford University Press
in the UK and certain other countries.

Published in the United States of America by
Oxford University Press
198 Madison Avenue, New York, NY 10016

© Oxford University Press 2013

All rights reserved. No part of this publication may be reproduced,
stored in a retrieval system, or transmitted, in any form or by any means, without the
prior permission in writing of Oxford University Press, or as expressly permitted by law,
by license, or under terms agreed with the appropriate reproduction rights organization.
Inquiries concerning reproduction outside the scope of the above should be sent
to the Rights Department, Oxford University Press, at the address above.

You must not circulate this work in any other form
and you must impose this same condition on any acquirer.

Library of Congress Cataloging-in-Publication Data
Graham, Daniel W.
Science before Socrates : Parmenides, Anaxagoras, and the
new astronomy / Daniel W. Graham.
pages cm
Includes bibliographical references.
ISBN 978-0-19-995978-5 (hardback : alk. paper)—ISBN (invalid) 978-0-19-995979-2 (UPDF)
1. Astronomy, Ancient—Greece. 2. Astronomy, Greek. 3. Parmenides.
4. Anaxagoras. 5. Science—Greece—History—To 1500. I. Title.
QB21.G73   2013
520.938—dc23   2012033087

1 3 5 7 9 8 6 4 2
Printed in the United States of America
on acid-free paper

*To my twin brother John, who shares
a love of antiquity and science*

CONTENTS

Preface   xi

Introduction: Cosmic Conjunctions   3

1. Looking for Science   7
    1.1. Unfounded Speculation   10
    1.2. Footnotes to Thales   16
    1.3. Footnotes to Pythagoras   18
    1.4. Science Without Knowledge   19
    1.5. History of Science Without History   24
    1.6. History of Science Without Science   25
    1.7. Old-Time History of Science   28

2. Azure Pastures: An Early Ionian Model   41
    2.1. Hesiod's Mythical Cosmography   41
    2.2. Ionian Theories   45
    2.3. The Meteorological Model   78

3. Borrowed Light: The Insights of Parmenides — 85
   3.1. Fifth-Century Advances — 85
   3.2. Three Insights: Heliophotism, Planetary Unification, Sphericity — 87
   3.3. The Power of a Model — 97
   3.4. Conjectures — 100
   3.5. Conceptual Advances — 104
   Conclusion — 107

4. Empire of the Sun: Implications of Heliophotism, and a New Model — 109
   4.1. Antiphraxis and Other Theoretical Implications — 111
   4.2. A New Physics — 122
   4.3. Anaxagoras' New Cosmology and Astronomy — 122
   4.4. The Lithic Model — 134

5. Darkened Suns and Falling Stars: Heaven-Sent Proofs — 137
   5.1. Lives of the Eminent Philosophers — 140
   5.2. Eclipses — 143
   5.3. The Meteor — 159
   5.4. The Comet — 165
   5.5. The Nile Floods — 170
   Conclusion: Theory and Evidence — 174

6. Lunar Revolutions: The Triumph of the New Astronomy — 177
   6.1. A Community Effort — 178
   6.2. Anaxagoras' Theory — 182
   6.3. Other Theories of the Fifth Century — 186
   6.4. Characteristics of the Lithic Model — 201
   6.5. The Doxography — 204

6.6. Plato's Heavenly Sphere	216
6.7. Aristotle's Paradigm	218
6.8. A Scientific Consensus	222
7. The Geometry of the Heavens	227
7.1. The Story of Early Greek Astronomy	228
7.2. Scientific Progress	237
7.3. Historical and Philosophical Significance	241
Appendix 1: Anaxagoras in the Historiography of Science	*247*
Appendix 2: Science and History	*255*
Bibliography	*261*
Index of Passages	*275*
Index	*281*

PREFACE

When the early Greeks looked up at the sky, they saw Helios, god of the sun, driving his blazing chariot, Selene, goddess of the moon, the Morning Star, the Evening Star, and countless lesser stars arranged in patterns depicting mythological characters and events. When we look up, we see a spherical star composed of hydrogen gas that our planet orbits, a spherical satellite composed of minerals and dust that orbits the earth, five visible planets that orbit the sun, and numerous pinpoints of light that are distant stars located many light-years from the sun.

At some point in early Greek history the mythological story got replaced, for intellectuals at least, by a naturalistic account strikingly like the modern one. It is the job of historians of science to tell us about how this and similar changes of conception took place. Yet historians have had surprisingly little to say about this transition, and have shown little curiosity about determining the when, where, why, how, and who behind the transition.

As a historian of philosophy and classicist, I got interested in the methodological advances made by early Greek philosophers in their pursuit of scientific explanations. I accepted axiomatically the *communis opinio* that these philosophers were good theoreticians but inept scientists—for they lacked the observational skills and empirical habits needed to confirm or disconfirm their theories. One of the major topics of early Greek inquiry was cosmology and astronomy. But so confident are most historians of their assessment that they do not even consider the possibility of astronomical advances among the Presocratic philosophers. Historians of astronomy, in particular, often begin their researches in the times of Plato; if they bother to mention early Greek astronomy (or cosmology, since they may refuse the honorific title of astronomy to early thinkers), they repeat the dismissive judgments of scholars from a half century ago.

In my own researches on methodology I was struck by the sudden advances in astronomical understanding of the early fifth century BC. Within the space of one generation, Greek philosophers went from having numerous ingenious but false theories of the heavens to sharing the same true (or nearly true) theory. How did this happen? Might the transition mark a genuine scientific advance, a discovery (as historians used to say before they started thinking of science as an invention)? I found that historians sometimes almost recognized a discoverer (though they disagreed who it was and how he made the discovery), but even when they did glimpse an advance, they found reasons to disparage or discount its scientific value.

I have written this book to try to fill a gap in the history of science. For, after taking a hard second look at early Greek cosmology, I became convinced that history has shortchanged the Presocratic philosophers. True, their theoretical imagination was better developed than their observational skills and their commitment to empirical testing. Yet for all their limitations, they realized something that no earlier civilization known to us had grasped, that the moon shines with reflected sunlight, and from that slender insight they worked out a geometry of the heavens that solved a series of cosmological and astronomical problems—so completely that the solutions have never been seriously questioned from the fifth century BC until the present. If this is so, we can push the history of scientific astronomy back roughly a century and a half and recognize a new paradigm of early science.

My own credentials as a historian of science are meager at best. But I have benefited from conversation, correspondence, and constructive criticism over a period of about twelve years to help me improve my ideas. In particular I wish to thank my colleagues Kent Nielsen, David Grandy, James Siebach, and astronomer Eric Hintz. I am also indebted to Andrew Gregory and Alexander Mourelatos for detailed critiques of the manuscript, as well as correspondence and meetings on scientific questions. I have also had the benefit of communications with Dmitri Panchenko and Dirk Couprie, both of whom take early astronomy seriously, and John T. Ramsey, an expert on ancient comets. My thanks to Justin Barney for manuscript preparation, and to Shelby Barney for help with graphics.

My wife Diana has been a constant support, and my daughter Sarah has read and commented on the manuscript.

SCIENCE BEFORE SOCRATES

Introduction

Cosmic Conjunctions

On February 17, 478 BC, shortly before noon, the sky in Athens began to grow dim. The late winter sun shone feebly and, seen in reflection or projected through leaves its bright disk narrowed to a crescent. Suddenly the sun went dark, except for a narrow ring of fire at the circumference. The people of Athens were witnessing a rare solar eclipse, this one an annular eclipse in which the sun is not completely obscured. Traditionally such an event was understood as a dire portent from heaven that struck observers with foreboding. But a new spirit of discovery was in the air, ever since, roughly a century earlier, the philosopher Thales had allegedly predicted a solar eclipse that was seen in Asia Minor. The present eclipse in Athens would usher in a new era, because for the first time someone looking up at it understood what was happening, recognized the event as a conjunction of heavenly bodies.

There had been a number of ingenious explanations of eclipses advanced by philosophers in the century since Thales' "prediction" of a solar eclipse around 585 BC, but none had even come close to a correct understanding. Now, a young refugee from Ionia, across the Aegean Sea from Athens, had a hypothesis about what could make the sun grow dark in a solar eclipse, and what could make the moon grow dark in a lunar eclipse. Standing in the rubble of Athens—for the great city had recently been demolished by Persian invaders—he made calculations that confirmed his hypothesis. Then he went down to the port city of Piraeus day after day for weeks or months and collected information from travelers. When he was done, he knew he had made a first-rate discovery.

The young man with the hypothesis was Anaxagoras of Clazomenae, who at the tender age of twenty-two solved a problem that no one in the history of the world had ever solved, and he proved his hypothesis empirically, on the basis of his—and others'—observations. Or so I will argue in the course of trying to explain the thought of this remarkable philosopher. Anaxagoras spent thirty years in Athens at the time of the Golden Age. His name became synonymous with esoteric knowledge, a kind of philosophical and scientific icon of his time, the nerdy intellectual, the Einstein of his age. But in time the book he wrote was lost, his theories were transmitted in a muddled form, and his contributions were obscured.

Some twelve years later another heavenly portent held Greece transfixed. A large comet blazed in the early morning skies in the east. Growing ever brighter and higher in the sky day by day as it approached the rising sun, its tail spread over the zenith of the sky. Shooting stars filled the sky for several nights. The comet disappeared briefly around July 18, 466, and then reappeared in the evening sky in the west as the sun set. It was visible for some seventy-five days in all. At some point after July 18, a ball of fire was seen falling in daylight in northern Greece. It hit the earth near Aegospotami with a thunderous explosion that threw debris far and wide. When the trembling inhabitants of the area approached the site they found a large smoking crater and a metallic stone the size of a wagon, of a burnt color.

What was striking was what did not happen to the stone of Aegospotami. It did not generate a mythic tale of a missile thrown by Zeus or one of the Olympians. Rather, rumor with its thousand tongues spread the word that the strange man from Clazomenae had anticipated the event, predicted it even. Sometime between the eclipse of 478 and the comet and meteor of 466 a profound shift had taken place in the Hellenic world. Instead of looking to soothsayers and bards for cosmic meaning, Greeks—some of them at least—were looking to philosophers. And sometimes these philosophers got things right. In the years between the eclipse and the meteor there emerged a new cosmology and a new astronomy. Already by the time of the eclipse there had been a multitude of cosmological theories proposed by natural philosophers, a number of naturalistic theories of how the world arose and operated. If the scope of these philosophers' inquiries was breathtaking, the results were disappointing. As one of the more circumspect of these thinkers put it,

> Now the plain truth no man has seen nor will any
> know concerning the gods and what I have said concerning all
>   things.
> For even if he should completely succeed in describing things as
>   they come to pass,
> nonetheless he himself does not know. (Xenophanes B34)

It is easy to theorize, the philosophers found, but hard to prove one's assertions—particularly concerning the remote phenomena of the heavens.

But something happened between the eclipse and the meteor. Suddenly some theoretical hypotheses began not only to make sense, but also to suggest empirical evidence that could be tested, and that was in fact tested and found to be right. The story of early philosophy and science has been told many times. But there remains a missing link, a moment of transition from theory to testing, from philosophy to science, from speculative cosmology to empirical astronomy that has been missed by even the most acute historians of philosophy and science. They have described the theories and events of the time and yet missed the significance. This study will be an essay in tracing the path from speculative philosophy to empirical science. The first empirical science attained by the Greeks, all agree, was astronomy. But received opinion puts its birth in the mid-fourth century BC at the earliest. By then, however, some of the most important astronomical discoveries had already been made, and the results known for over a century. This study will attempt to set the clock back on empirical science, and to give credit where credit is due, to some very unlikely founders.

This study will also have something to say about the theory and practice of science itself, especially in light of recent attacks on the integrity of science and the objectivity of its history. The ultimate answer to such attacks is perhaps not more theory—though theory is important—but historical evidence showing that science can take care of itself. And if the short view of a few decades or even several centuries will not offer a convincing answer, perhaps the long perspective of two and a half millennia will.

Chapter 1

Looking for Science

> [Helios] shines on mortal men and immortal gods as he drives his horses; terribly he gazes with his eyes from under his golden helmet, and bright beams from his person blaze dazzling, and about his temples the glittering locks of his head gracefully surround his far-shining face. ... There having halted his gold-yoked chariot and horses, he miraculously conducts them through heaven Oceanward.
> —*Homeric Hymn to Helios* 8–16[1]

When did a science of the heavens begin? This question is impossible to answer. We know from archaeology that massive stone structures such as Stonehenge were aligned with celestial phenomena thousands of years BC. We know from cuneiform texts that the Babylonians were recording minute observations of stars and planets more than seventeen hundred years BC. Chinese and Egyptian records of the heavens reach back into antiquity. Phoenician sailors learned how to use stars to navigate the open ocean. Yet much of early stargazing seems to have been conducted for the sake of the practical needs of a usable calendar, on the one hand, which would allow farmers to know when to plow and plant or sailors and fishermen when to go to sea; and for the deciphering of omens, on the other hand, to know what benefits or calamities to expect. In poetry and myth, the sun and moon were personified as deities. Shining Helios drove a chariot across the sky (see epigraph)

---

1. My own translation, as are all translations except as otherwise noted.

each day to illuminate the earth. The Greek mythological view of the heavens was not notably different from those of neighboring cultures.

In the sixth century BC, a few innovative Greek thinkers looked up to the heavens for a different purpose. Where their contemporaries saw Helios, the sun god in his chariot, Selene, the moon goddess, and other divine beings, the philosophers saw natural formations that were part of the cosmos, the ordered world. They produced numerous competing models of the heavens, but they shared a belief that what was up there was not the host of heaven, but some set of natural bodies or phenomena that were either continuous with or at least not in principle different from the clouds and events of the atmosphere around us—which themselves were not deities or divine manifestations.

Somehow from the theories of the Greek thinkers and from the observations collected by their neighbors, especially the Babylonians, arose the advanced astronomical science of late antiquity. The great synthesizer, Claudius Ptolemy, wrote in the tradition of Greek mathematics and science, but drew on the extensive chronicles and compendia of Middle Eastern sources. It is a well-accepted truth that sometime between the time of the first Greek speculations and the *Almagest* of Ptolemy (c. AD 150), a scientific astronomy arose among the Greeks. This does not mean, however, that the Greeks were the only ones to develop a scientific astronomy. Indeed, recent researchers have shown that starting centuries earlier than the Greeks, the Babylonian tradition produced a science of the heavens that rivaled and in many respects excelled its Greek counterpart, employing both extended observations and mathematical models. And yet, for all its limitations, Greek astronomy made unique contributions to a knowledge of the heavens in its own way. "Astronomy is often considered the zenith of the 'exact sciences' in antiquity" (Rihll 1999:62), and "The greatest achievements of early Greek science lie in astronomy" (Lloyd 1970:80). In astronomy, unlike other Greek scientific studies such as biology, there was a steady and progressive development over a long period of time, with major contributions from Middle Eastern science, in a tradition that continued through medieval to early modern times. Accordingly, astronomy represents the paradigmatic Greek science and offers a window into the origins of Western science.

It is the consensus of historians of science that scientific astronomy did not arise among the Greeks before Eudoxus in the fourth century BC, and perhaps even later. Apparently Plato, in whose Academy Eudoxus worked, posed general questions that helped define and organize scientific attitudes towards the heavens.[2] And most scholars recognize Meton and Euctemon, who worked in Athens around 430 BC, as the first Greeks to make important advances in empirical astronomy.[3] The former recommended correlating nineteen years with 235 lunar months; the latter measured the length of the seasons and came to the surprising result that they were not equal in length. Their practical studies in preparing a workable calendar provided the foundations of scientific astronomy.

Meanwhile, the early thinkers, whom we call Presocratic philosophers, are given credit for their bold hypotheses and physical models of the heavens, but dismissed as unscientific because of their lack of a method for testing their theories. I have in the past accepted this judgment, even while working to prove that the Presocratics, especially those of the Ionian tradition, were more systematically scientific in their attitudes and practices than has often been recognized.[4] In the process of studying the Ionians, I have come slowly and somewhat reluctantly to the conclusion that they not only brought scientific attitudes to the world, but made genuine scientific progress of the sort that deserves to be identified as a set of scientific discoveries,

---

2. See Vlastos 1975.
3. Dicks 1970:87–89; Lloyd 1979:171–73. But Bowen and Goldstein (1988) have argued plausibly that Meton's study did not depend on precise observations at all, but rather on schematic synchronisms borrowed from Babylonian astronomy. See below, ch. 7.1.
4. See especially Graham 2006.

particularly in the field of astronomy. This book is an attempt to make the case that before Plato, Eudoxus, even Meton and Euctemon, and before Socrates, the great turning point of philosophical method, scientific advances were made in astronomy, and made precisely by those figures experts say could not have made them.

This brings us to a confrontation with the standard views on early Greek astronomy. They are well researched, well established, and have been built up over several centuries. I begin with the dominant view, a pessimistic one, which argues strongly against any important contributions from the Presocratics.

1.1 UNFOUNDED SPECULATION

Students of Presocratic philosophy typically praise the early thinkers for their naturalistic understanding of the world. These thinkers propose brilliant hypotheses about the nature of the world and its phenomena. Thales says that all is water, Anaximenes that all is air, Heraclitus that all is fire. Anaximander says that the heavenly bodies are like great wheels, Anaximenes that they are like leaves, Heraclitus that they are like bowls, Xenophanes that they are like coals, or clouds, Anaxagoras that they are like stones. On the basis of their hypotheses they explain other phenomena such as day and night, the phases of the moon, eclipses, comets, shooting stars. The problem is that for all their fecundity in theorizing, the Presocratics fail to provide any way to test their theories. So in the end every theory is equally plausible and equally unprovable and unproved. We seem to have a rich tradition of theorizing without a corresponding practice of testing and verification.

Some authoritative statements of this view are the following:

> In sum, the main preoccupation of the later Presocratic philosophers was with the problem of change. . . . Empedocles,

LOOKING FOR SCIENCE

Anaxagoras, Leucippus and Democritus were chiefly engaged not in programmes of research, but in discussions of a highly abstract nature in which what counted was not the empirical data that could be adduced in support of theory, so much as the economy and consistency of the arguments on which it was based. (Lloyd 1970:49)[5]

[The ideas of the Presocratics] are the dream children of the speculative thinker in the study intoxicated by the novelty and daring of the new intellectual atmosphere and intent on applying the new methods of thought... in the widest possible field. [The Presocratics] were *not*, however, primarily scientists, much less astronomers, and observation of actual celestial phenomena seem to have played a relatively minor role in their thinking. (Dicks 1970:60)

We do not know of a single fact observable by the *physiologoi* [the Presocratic students of nature] with the means at their disposal which could "upset" any major theory of theirs in the domain of terrestrial physics. (Vlastos 1975:87)

In coming to understand and explain [nature], they rarely used careful observational data, or experiments, in support of their claims. Nevertheless, the problems that the pre-Socratic philosophers identified, and with which they grappled, largely by abstract, rational arguments, formed the basis of natural philosophy as it would be shaped in the fourth century BC by Aristotle. (Grant 2007:18)

Undoubtedly many cases of Presocratic theorizing exemplify the criticisms that are raised by modern critics. The question, however, is whether all cases do.

---

5. On Presocratic empirical research, see Lloyd 1979:139–45; specifically on astronomy, 169–73.

For the moment I would like to focus on the conclusions of one prominent expert on early astronomy, D. R. Dicks. After more than forty years, his book, *Early Greek Astronomy to Aristotle* (1970), remains the standard authority in the field. It follows closely the developments of early Greek astronomy in light of a rigorous understanding of the phenomena. Although there have been criticisms of the book,[6] they have had little impact, and no comparable book has emerged to take its place. The brief against the Presocratics as scientific students of the heavens is invariably the one mentioned by Dicks. They are too speculative, and not sufficiently empirical. Dicks gives them credit, as do virtually all historians of science, for developing a method that had the potential to solve the problems of science: "by abandoning mythological traditions and subjecting external phenomena to a process of logical reasoning, untrammelled by religious dogma, and even by investigating the actual processes of thought, they opened up a whole new field of knowledge which is virtually inexhaustible" (60). To put the point in the discourse of science, they were good at advancing hypotheses and at discussing them critically in an open forum, but bad at testing them against observational data.

Failures of observation are not, however, the only problem Dicks sees. In some sense, surprisingly, their theories were not abstract enough. Their theories "are all *qualitative* in conception" (60, italics in original). The theories lacked a mathematical basis that could provide a truly scientific view of the heavens. Dicks seems to think that a close observation of risings, settings, and the like would actually provide the mathematical background needed for a robustly scientific conception of the heavens. Philolaus' innovative scheme in which the sun, moon, and planets circle a central fire is an example of bold thinking without sufficient empirical grounding: "it is very much the product of the study and bears little relation to the facts of actual observation (e.g., it

---

6. See especially Kahn 1970.

takes no account of the latitudinal and longitudinal variations of the planetary bodies)" (70). In general, scholars "grossly exaggerate the mathematical and scientific content of early Greek thought."[7] The claims of Presocratic theories are too vague, too imprecise, too unmathematical, and hence not testable by reference to empirical data.

Thus, in general the Presocratics brought a scientific attitude to the study of astronomy. They proposed general theories, but these were qualitative rather than quantitative in character. The Presocratics were indifferent observers and hence unable to test their theories. Overall, they were too speculative and insufficiently empirical in their approach to explaining heavenly phenomena. They glimpsed from afar the proper method for understanding the heavens, but they failed to enter into the promised land of science.

Consequently, many scholars hold that the Presocratics should be viewed as cosmologists or natural philosophers rather than as scientific astronomers. Studies of "early Greek astronomy" typically begin with Eudoxus or even later figures.[8] Histories and textbooks treat Presocratic figures as having interesting things to say about the heavens but making no real contributions to an understanding of the phenomena.

It is useful to recognize an evolution in problems and methods of astronomy. There seem to be stages of development in which practices within one period are similar but differ from those of another.

---

7. Dicks 1966:38. Dicks also claims that much of the knowledge attributed to early thinkers was impossible at that stage of understanding, including an awareness of planets, equinoxes, and the lengths of seasons and years (1966).

8. Kalfas 1990:121. Similarly Clagett 1963, ch. 7; Wilson 1997:24–26; these works pay lip service to earlier figures, but exclude them from important discussions. Neugebauer 1972, "On Some Aspects of Early Greek Astronomy," focuses on Hipparchus as an exemplar of "early" Greek astronomy (2nd c. BC). Rihll 1999 does not mention Anaxagoras in the context of astronomy (ch. 4). North 1995:66, 78 allots Anaxagoras two sentences, one giving cautious recognition to his account of solar (but not lunar) eclipses. Grant 2007 does not even mention Anaxagoras. See below, Appendix 1, for a more extended review of histories of astronomy.

Roughly, one can say that different paradigms or research programs dominate different periods. In "A New View of Early Greek Astronomy," Goldstein and Bowen (1983) present a helpful periodization of the field for the standard view. "We contend," they say,

> that Greek astronomy falls into distinct phases, that the earliest phase extends from the eighth century BC to the time of Plato, and that Eudoxus is the starting point of the second phase. This division is based on philological considerations and on the observation that, in the earlier period, the motives and problems of Greek astronomy proper differed from those of early cosmology and that any scientific concern with planetary motion was a characteristic of the second period. (330–31)

This general conception is broadly consistent with the view of Dicks and also of that of Vassilis Kalfas:[9] Eudoxus was the watershed who turned astronomy into a mathematical study and in that sense, a science proper.

According to Goldstein and Bowen, "Greek astronomy, as the very name implies, began as the organization of the fixed stars into constellations. The purpose was to construct a calendar by correlating dates and weather phenomena with the risings and settings of the fixed stars or constellations" (331). This work was gathered in a *parapēgma*, perhaps originally by Meton. (The *parapēgma* was a calendar on a board with a hole under each date; a peg was placed in the hole of the current date.)[10] The determinations were crude and

---

9. Kalfas stresses Eudoxus as the starting point of scientific astronomy, whereas Goldstein and Bowen allow for an earlier, non-mathematical stage; but all agree as to the importance of Eudoxus in introducing mathematical models.
10. See Lehoux 2007, who warns that we do not actually know how early these calendars were constructed, and Taub 2003, ch. 2. For Meton's project as focused on a *parapēgma*, see Bowen and Goldstein 1988.

made no mention of planetary motions. The peg-board calendar was needed at least in part to mark times for planting, reaping, sailing, and so on, for Greek civic years based on lunar months were virtually always out of synch with the solar year.[11] "Since the accounts of celestial motions found in the tradition of cosmological speculation, a tradition which was independent of Greek astronomy in its earliest phase, are ... tangential to the strictly scientific treatment of the same phenomena, it appears that all aspects of astronomy prior to Eudoxus concern the calendar and calendaric cycles" (332).

On this view, the unfounded speculations of the philosophers really had nothing to do with astronomy and can largely be ignored. Only when Eudoxus combines mathematical modeling with the cosmological speculations of the Pythagoreans and Plato do we get a confluence of philosophy and science (332 ff.). His "synthesis" of cosmology and measurement is responsible for the sophisticated scientific astronomy of the second period (340).[12] Before Eudoxus, philosophy is largely irrelevant to astronomy.

This dismissive attitude toward the Presocratics is pervasive among contemporary scholars, but hardly new. Indeed, it reaches back to the very end of the Presocratic period. Xenophon says about Socrates, "he wondered why it was not plain to [cosmologists] that it was impossible for men to discover [*heurein*] these things, inasmuch as even the most advanced theorists did not agree with each other" (*Memorabilia* 1.1.13). This objection seems to point to the lack of evidence as seen from the failure of consensus. It appears that from the time of Socrates to the present, Presocratic theories have been rejected as insufficiently scientific. The author of the Hippocratic *Ancient Medicine* says, "I assert that [medicine] has no need of a vain

---

11. Twelve lunar months are about eleven days short of a solar (tropical) year. When the months became too displaced from their season, Greek cities (each with its own system and nomenclature) intercalated a whole extra month, whenever its magistrates decided to do so.
12. For a look at Eudoxus' shortcomings, see below, ch. 7.1.

hypothesis as to remote and bewildering phenomena, such as those in the sky or under the earth, concerning which it is necessary, if one is to explain them, to use a hypothesis. But if [13] one should explain or claim to understand what these things are, it would not be clear either to the speaker or to the audience whether his statements were true or not. For there is nothing against which to test one's knowledge" (ch. 1). Here again the point is the lack of evidence, and the impossibility of testing a theory against fact.

It appears, then, that there has been a presumption against Presocratic astronomy that goes back at least to the fourth century B.C. and continues unabated to the present. The Presocratics are guilty of airy speculation unsupported by an adequate basis of empirical data that could prove or disprove their theories. I shall call this the Unfounded Speculation view, and regard it as the standard response to Presocratic science and in particular astronomy or (if that word seems too honorific) cosmology. "Rational as Presocratic cosmology may have been ... it remained in some ways little more than *unfounded speculation*" (Hetherington 1993:65, my italics). This view embodies extreme pessimism about the outcomes of early researches.

On the other hand there is an extremely optimistic counterpart.

1.2 FOOTNOTES TO THALES

A minority of scholars has found in ancient testimonies the grounds for a very different reading of early Greek astronomy. There is fairly good early evidence that Thales, the first philosopher-scientist, "predicted" a solar eclipse.[14] The eclipse is plausibly connected with one verifiable by modern science that was visible in Asia Minor in 585

---

13. Reading ἂ εἰ for ἀεί with Littré.
14. Herodotus 1.74.2 = DK 11A5; Pliny *Natural History* 2.53.

B.C.[15] Thus at the very beginnings of philosophy and science, we find a gifted thinker who was able to grasp the complex relationships between the heavenly bodies and to explain and predict an event that happens rarely and that had up until that time been regarded as a divine omen or portent rather than a natural occurrence.

If this is right, then Presocratic philosophy and science began with a stunning breakthrough that served as a paradigm for all later researches. As one prominent defender of the view puts it, "I assume that Thales had indeed some acquaintance with Mesopotamian astronomy and was familiar with the notion of eclipse cycles" (Panchenko 1994:279). "As a respectable Greek from the most important Ionian city, Miletus, Thales could easily have met at the court of Pharaoh Necho (610–595 BC) various people from Assyria or Phoenicia" (279). Drawing on the advances of Babylonian and Egyptian science, Thales was able to use the Exeligmos eclipse cycle to anticipate an eclipse.[16]

Furthermore, on this view, Thales understood how the sun illuminates the moon; how when the moon is in precise conjunction with the sun, it occults the sun.[17] Aided by Babylonian tables he was able to determine when a precise conjunction would occur, or at least was likely, and accordingly predicted a solar eclipse, to the astonishment of his contemporaries.[18]

---

15. Recent computer models eliminate other candidates: Stephenson and Fatoohi 1997.
16. In the 1994 article, Panchenko favors the year 582/1 BC, in which two solar eclipses occurred. Even those who are more cautious about Thales' knowledge believe that much Greek astronomy came from the Babylonians: "it is often just as likely that the Presocratic philosophers derived their knowledge [of astronomy] directly or indirectly from the East as that they made the discoveries independently" (Lloyd 1970:81). But it is not likely that the Greeks borrowed astronomical theories from the Babylonians, since the latters' astronomy was computational in character: Lloyd 1991c:294.
17. Panchenko 1996:121–22; Panchenko 2002:332, 334. (The former is in Russian with an English summary; I have read only the summary.) Cf. Rossetti 2011.
18. See Panchenko 1994; O'Grady 2002:126–46. They give different dates for the eclipse; for Panchenko, see n.16; O'Grady accepts the eclipse of 585.

After Thales there was a period of relative stagnation when new but mistaken theories were tried.[19] But by the early fifth century, the principles of Thales were firmly established and astronomy could continue to progress toward the mathematical theories of the fourth century. Thales had also introduced geometry to Greece from Egypt and made major advances in that study, which prepared the way for the mathematical description of heavenly motions.

Thus, there was a sudden burst of creative energy and scientific discovery at the very beginning of philosophical and scientific study of the world. On this view, the Greeks are deeply indebted to their neighbors for scientific knowledge: to the Babylonians for astronomy and to the Egyptians for geometry. Thales was the great innovator, or at least the great transmitter of knowledge, and the histories of philosophy and astronomy are a series of Footnotes to Thales.

1.3 FOOTNOTES TO PYTHAGORAS

Pythagoras (ca. 570–490 BC) emigrated from the island of Samos on the Ionian coast to Croton in southern Italy. There he founded a religious society that lasted well into the fourth century BC. He apparently did not leave any writings to posterity, but a strong tradition allowed for the transmission of ideas. Besides his doctrine of transmigration of souls, which is attested by early sources, he is associated with a doctrine of numbers and a cosmology by later sources. One of his alleged successes is to recognize that the moon gets its light from the sun.[20] More vaguely, the Pythagoreans are said to give the correct account of solar eclipses and "some" Pythagoreans the correct account of lunar

---

19. But Panchenko 1996:122–23 holds that Anaximander, and possibly Xenophanes and Heraclitus had a correct understanding of eclipses. On this, see ch. 2 below.
20. Aetius 2.28.5.

eclipses.[21] Scholars have often assumed that either Pythagoras himself or some early members of his society or "school" developed a mathematical approach to the world and a mathematical astronomy that correctly accounted for the moon's light and eclipses in the sixth century BC. Other figures in the fifth century adopted these theories and incorporated them into their cosmologies. Thus Greek astronomy is a series of Footnotes to Pythagoras.[22]

1.4 SCIENCE WITHOUT KNOWLEDGE

Interpretations of early Greek astronomy can be plotted between these extremes of pessimism and optimism, the Unfounded Speculation view and the Footnotes to Thales or Pythagoras view, all of which go back more than a century. Both of them are based in some measure on a classic understanding of science as a progressive discovery of laws of nature. In the last half century, however, this understanding has been called into question by philosophers of science. Most influential has been the work of the philosopher of science Thomas Kuhn, who in his influential study *The Structure of Scientific Revolutions* has argued that science does not progress in a steady way (1962). Scientists work out a "paradigm" or example of successful scientific explanation that later scientists follow. This establishes a pattern of "normal science" in which standard methods are applied to

---

21. Aetius 2.24.6 (§2 Mansfeld and Runia); Aetius 2.29.4.
22. Recent Rovelli 2011 has argued that Anaximander is the first scientist. Anaximander has a much better claim than either Thales or Pythagoras, and I would agree that scientific *philosophy* is a series of footnotes to Anaximander, and in that sense Western science is deeply indebted to him. But in a strict sense of science, he does not seem to qualify as a scientist. For the contributions of Anaximander, see Kahn 1960; Couprie in Couprie, Hahn, and Naddaf 2003; Graham 2006, ch. 2; Couprie 2011, Part 2. On the concept of science, see below, Appendix 2. As the view that science is a series of footnotes to Anaximander is new and not widespread, I shall not address it formally; but my argument in ch. 2 below and elsewhere implicitly shows why Anaximander should not count as a scientist.

solve outstanding questions. As time goes on, however, "anomalies" pile up, leading to a period of "crisis" followed by a revolutionary episode in which a new approach is proposed, which eventually provides a new paradigm for a different kind of scientific approach. During the time of crisis two or more approaches compete. They are so different as to be "incommensurable," admitting of no simple algorithm or rational decision procedure as to which is better. Ultimately, a new generation of scientists opts for a new paradigm, which then replaces the old, not by any rational process but simply by winning the hearts of a new generation. In many ways, Kuhn's analysis provides a helpful account of stages in scientific research and a recognition of a historical dimension.[23] Yet it ultimately views major scientific change as a kind of social-political process independent of rational norms and unconstrained by any generally accepted evidence.

On this "constructivist" model, there is a kind of relative progress within normal science, and a historical succession of scientific approaches, but no objective progress in science, simply because there are no independent criteria capable of judging between the claims of competing systems. Kuhn built his theory of scientific change on a study of historical developments in science that seemed to him to undercut the classical account of scientific progress. Subsequently, other theorists such as Imre Lakatos (1970) and Larry Laudan (1977) have suggested ways in which a more robust notion of progress within a larger "research program" or "research tradition" is possible; yet they still fall short of endorsing a notion of absolute progress in understanding the world or getting at the truth of the matter. Consequently we may say that all of these thinkers advocate a Science Without Progress. Without progress, there seems to be no way to validate science as an activity that stands out as especially effective in

---

23. I have used it to clarify issues in the development of Presocratic methodology, in Graham 2006, e.g., 91–2.

finding out the secrets of the world. Furthermore, it becomes dubious whether science, given its dependence on contingent paradigms, can ever arrive at truths about the world.

Indeed, the whole notion of truth as a goal of science has come under attack. Perhaps the most outspoken defender of truth as a goal of science, Karl Popper (see 1968) turned out not to have any way to approach truth, but only a *via negativa* to eliminate error. For him truth is an ideal scientists somehow seek, but it plays no role in scientific research. As for progress, the scientific method helps to eliminate errors, but does not lead to any linear notion of advancement.

Instrumentalist philosophers of science fully acknowledge the success of science in making predictions and allowing control of physical processes, but deny that this has any connection to truth. According to them, science provides only a convenient way of calculating observable events from observable conditions. In principle, the particular theory we use could be replaced by other equivalent theories that do not commit one to the theoretical entities of the first; therefore the theoretical entities mentioned in a given theory are not themselves essential to explanation, and nothing is proved about the non-observable structure of the world. Instrumentalists clearly advocate a Science Without Truth view.[24]

The view that constructivism, instrumentalism, and similar theories reject is scientific realism, the position that the existence of the theoretical entities of well-confirmed theories is supported by tests of the theory. According to scientific realism:

1. Theoretical terms in scientific theories (i.e., nonobservational terms) should be thought of as putatively referring expressions....

---

[24] Ultimately, constructivists too reject truth as a basis of science; but the point is more prominent with the instrumentalists, who allow for progress but reject truth as a source of the progress.

2. Scientific theories, interpreted [according to (1)], are confirmable and in fact are often confirmed as approximately true by ordinary scientific evidence interpreted in accordance with ordinary methodological standards.
3. The historical progress of mature sciences is largely a matter of successively more accurate approximations to the truth about both observable and unobservable phenomena. Later theories typically build upon the ... knowledge embodied in previous theories.
4. The reality which our scientific theories describe is largely independent of our thoughts or theoretical commitments. (Boyd 1984:41–42)

On this view theoretical terms ideally refer to realities; scientific theories make testable truth claims about the world, which science tests experimentally; on the basis of improving approximations to the truth, theories become progressively better at describing and predicting events; and science is an objective rather than a subjective procedure for finding out about the world.

The most powerful argument advanced against scientific realism is "the pessimistic induction" (or "meta-induction").[25] According to this argument, all previous scientific theories have been superseded by later theories. Therefore, all previous scientific theories have been proved false. But then it is likely that all our current scientific theories will be superseded in turn. Hence, it is likely that all our current scientific theories are false. On this view, the history of science offers a graveyard of theories and prefigures the doom of every scientific theory we hold dear.

Based in part on anti-realist accounts of science, cultural critics under the banners of post-modernism and post-structuralism also reject the

---

25. See Putnam 1978:24–25, focusing on the reference of terms in scientific theories. On the importance of this argument, see Devitt 2010:86.

claims of science to represent secure knowledge that is somehow better or more reliable than what people can arrive at by ordinary processes of negotiation and consensus. If this is so, then the whole project of the history of science, to track and illuminate the path from ignorance to knowledge, and from random discoveries to reliable methods, is an illusion. The history of science is no more than a certain kind of social history or sociology of a field that in the past has made inflated and unjustifiable claims to truth and social utility. For many specialists, even in the history of science, the history of science has increasingly become a sociology of science in which one studies how scientific ideas are arrived at, approved by a community of power brokers, and promulgated by their authority. Science has no timeless truths to establish, but only constructions that a powerful elite is able to sanction and promote to a suitably awed lay audience.[26] We arrive at a view of Science Without Objectivity.

The issues raised by recent philosophy of science go far beyond the scope of this study. Yet they cannot simply be ignored either. For if the Science Without Progress, Science Without Truth, and Science Without Objectivity views are correct, the whole question of who first discovered what in astronomy is misconceived. Each culture and even each scientific subculture plays its own game by its own rules. Scientific "discoveries" are at best advances made relative to a paradigm valid only within a limited tradition. There are, one might say, histories of scientific projects, but there is no transcendent unified history of science to be told. Instrumentalists can say that there are better ways of organizing data to make predictions, but these are not discoveries so much as improvements in computational methods. Questions of historical priority, of first discoverer of some principle, what the Greeks called the *prōtos heuretēs*,[27] are a grand waste of time, the sign of an obsolete methodology and bankrupt theory.

---

26. See Norris 1997.
27. See Kleingünther 1933, Zhmud 2006, ch. 1.

I shall lump together the several anti-realist, relativist, and constructivist views sketched here as the Science Without Knowledge approach, an approach that tends to demote the history of science, that is, to assimilate it to a history of ideas or practices that is not in principle different from any other intellectual or social history except in the pretensions of its practitioners.

1.5 HISTORY OF SCIENCE WITHOUT HISTORY

As if the objectivity of science was not problematic enough there is a further problem: Is history objective? At least with science, one can speak of experiments with intersubjectively observable results. And these experiments are in principle repeatable (even if experiments from long ago were done with different equipment and under different conditions, and even if some skeptics dispute the possibility of objective observations). But history by its nature consists of a unique series of events that are apparently contingent; that is, dependent on particular circumstances and participants in a peculiar situation, and hence in principle never repeatable. Thus if a knowledge of science is problematic, a knowledge of history is even more so. There are also ontological problems with history: Does the past even exist? If so, in what way? And there are epistemological questions peculiar to history. For instance, it has traditionally been held (since the time of Plato and Aristotle) that scientific knowledge is of universal truths. But history yields only particular truths. So how can there be rigorous knowledge of history?

First, a few distinctions. "History" is ambiguous between (a) the events of the past and (b) a critical representation of (a) as expressed in narratives, explanations, and arguments. (For instance, (a) "Throughout history, people have believed . . ."; (b) "I am writing a history of the Peloponnesian War.") History in the second sense is also called "historiography." But "historiography" itself is ambiguous between (a) a critical representation of the past and (b) a "meta-discipline" consisting of a second-order study of historiography in the first sense.[28] This latter sense of historiography can be called more properly "philosophy of historiography" and seen as a sub-discipline of epistemology.[29] The present chapter will be in part an exercise in the philosophy of historiography, while most of the present book will be an essay in the historiography of ancient science.

Now, does historiography in general and historiography of science in particular produce objective knowledge of the past? Positivist science assumed that there were pre-packaged facts about the world to be explained by their being deduced from a law-like statement. Similarly, positivist history assumed that there were facts about history (a) to be explained. In this scenario, the history of science is a relatively easy inquiry to pursue and immediately yields objective knowledge. The demise of positivism has brought an end to this naive picture. Some philosophers have gone to the opposite extreme of believing that the relevant facts are completely determined by theory without any empirical constraints. In this new scenario, the history of science is also a relatively easy inquiry to pursue, but yields no objective knowledge, only subjective satisfaction. And we arrive at a view of History of Science Without (objective) History.

1.6 HISTORY OF SCIENCE WITHOUT SCIENCE

In the last two sections I have noted far-reaching criticisms of traditional history and philosophy of science that threaten to nullify traditional inquiries into the origins and development of science. Yet a

---

28. Kragh 1987:20–21; Tucker 2004:1.
29. Tucker 2004:2.

new approach has arisen that promises a middle way between the old-fashioned idolatry of science and scientific method, and the postmodern reduction of every enterprise to an all-too-human power-structure. In 1992 the history of science journal *Isis* (83.4) presented a collection of papers on the history of science, "The Cultures of Ancient Science: Some Historical Reflections," by distinguished scholars who specialize precisely in ancient science, and who provide a kind of illustration and manifesto of the new approach.[30] The papers herald "a turning point in the historiography of science" (548). In the introductory essay, Francesca Rochberg acknowledges the impact of researches in ancient traditions other than the Greek one, and observes that "the non-and pre-Greek, non-and pre-Western material fundamentally challenges our old definitions of science, our understanding of its development, and consequently the way in which we should reconstruct and write its history" (548). Noting the influence of Kuhn's philosophy of science and the concomitant danger that all talk across paradigms or cultures might come to be viewed as incommensurable, she sees in the essays of the collection "a middle ground" (549) between a monolithic march-of-science account and a barren historical relativism. The papers of the collection "implicitly criticize an earlier historiography of science that presupposes progress as a characteristic of science and objectivity as a characteristic of its historians" (551).[31]

In his essay, David Pingree contrasts a virtuous Philhellenism with a vicious Hellenophilia (554–55). It quickly becomes clear that

---

30. There are papers by Francesca Rochberg, David Pingree, G. E. R. Lloyd, Heinrich von Staden, and Martin Bernal. For further information, see under the individual authors in the Bibliography of this volume. (Unlike the others, Bernal is not a historian of science, but addresses a scientific topic.)
31. And yet I cannot help but wonder what science is without progress, and what history is without objectivity—even if progress is often problematic and objectivity more of an ideal than a reality. On historical objectivity, see above, ch. 1.5.

Hellenophilia is the *bête noire* of the new historiography of ancient science:[32]

> a Hellenophile suffers from a form of madness that blinds him or her to the historical truth and creates in the imagination the idea that one of several false propositions is true. The first of these is that the Greeks invented science; the second is that they discovered a way to truth, the scientific method, that we are now successfully following; the third is that the only real sciences are those that began in Greece; and the fourth . . . is that the true definition of science is just that which scientists happen to be doing now. . . .[33] (555)

Each of these propositions "distorts the history of science in two ways: passively, it limits the phenomena that the historian is willing or able to examine; actively, it perverts understanding of both Western sciences, from the Greeks till now, and of non-Western sciences" (557).

What, then, is science for the new historians of ancient science? "I would offer this as the simplest, broadest, and most useful [definition]: science is a systematic explanation of perceived *or imaginary* phenomena, or else is based on such an explanation" (559, my italics). Pingree is in fact keen to include among his sciences astrology, magic, and divination. He is singularly broad-minded and generous to the underappreciated cultures of the world. Now, it appears, we are free to include mythology, religion, and just about any practice that purports to explain anything under the rubric of science.

---

32. The term "new historiography of science" is sometimes used to refer to the kind of historiography championed by Koyré and Kuhn (see Pinto de Oliveira 2012). Of course "new" is "token-reflexive" and tends to attach to the latest thing. Here I mean to pick out the specific program sketched in *Isis* 1992.
33. *Sic*; surely the extension of the term is not its definition.

Yet there is a danger in making the history of science a study that immerses itself completely in the historical situation, accepting only the categories of the target culture. If every science is a body of knowledge peculiar to a unique culture, and if there is no scientific progress, and if every culture has only its own unique institutions, then everything is indeed relevant to science, but science itself seems to have dwindled into just another topic of Cultural Studies. We avoid Hellenophilia only to embrace Barbarophilia. If, indeed, all the features we find admirable in science are discounted or even rejected by the new historiography—continuity, progress, objectivity—and "affinity" or similarity between ancient and modern exemplars is considered a vice,[34] then we shall find that the "science" of ancient science and the "science" of modern science are equivocal terms. We potentially arrive at a History of Science Without Science.

1.7 OLD-TIME HISTORY OF SCIENCE

There are, then, reasons to despair of a resolution to the problem of early Greek science. On the one hand a gloomy view holds that however brilliant the insights of the early Greek thinkers, they had no way of verifying their claims and of adjudicating between one astronomical theory and another. On the other hand, a grandly optimistic view holds that the very first philosopher-scientist-*Wunderkind* made all the important discoveries at one stroke and left little for his successors but belatedly to come to appreciate his sudden advances; alternatively a religious guru did this. Further, influential theories in the philosophy of science warn that science has nothing to do with truth

---

34. See Rochberg 1992:551, 552, Pingree 1992:555, von Staden 1992:582, 583, 586 for these vices.

and perhaps scientific objectivity and progress are illusions as well, and to try to chronicle the absolute advances of science is a waste of time. The new history of science seems to dismiss ancient Greek astronomy precisely because it has affinities and historical connections with modern scientific astronomy.

1.7.1 Substantive questions

With proponents of the first three views I agree that the question of scientific progress in early Greek astronomy is worth pursuing. On points of detail I disagree with all. Against the Unfounded Speculation view I shall try to show in subsequent chapters that there is adequate historical evidence for a progressive articulation of a theory of the heavens that is supported by empirical evidence—evidence the ancient theorists in fact drew on to justify their hypotheses. Further, I shall show that they were successful in convincing their contemporaries and successors of their theories, on the basis of the available evidence. This will be a piece of old-time history of science in which a scientific hypothesis is vindicated. Along the way I shall show why the Footnotes to Thales view and the Footnotes to Pythagoras view are dubious: there is no supporting evidence to confirm a similar level of understanding for Thales or his early successors. There will be a hero in this story, in fact two heroes, but they will not be Thales or Pythagoras.

1.7.2 Methodology: Science Without Knowledge

As to the Science Without Knowledge approach, I have a preliminary argument to offer and a story to tell. First, the argument. It is a fact beyond rational dispute that science, at least modern science, is consistently and demonstrably successful in an often spectacular way. Indeed, that it is progressing in its ability to predict and control physical events is incontrovertible. Science can account for phenomena from the subatomic level to the extent and lifespan of the universe. It has mapped the human genome, beheld the early history of the universe, explained the movements of continents, and split the atom into ever-smaller components. Applied science has sent humans to the moon, space probes to the outer reaches of the solar system and beyond, is engineering structures at the nano scale, cloning organisms, modifying genes, and examining brain processes in real time. While scientists are themselves products of a culture and a time, its results are not culture-relative in any meaningful sense. Though we may respect cultures (including that of early Greece) that say otherwise, the earth is (roughly) spherical. The sun shines by a process of nuclear fusion. However much we may wish it were otherwise, humans have artificially created nuclear fission and nuclear fusion, and have the standing capacity to unleash destructive forms of this energy. Science works. Indeed it has its political, social, and psychological dimensions, but it still works as a way of explaining, predicting, and controlling nature. Our job as philosophers is to accept this fact, and see what it implies for the way things are. This was, in essence, Kant's reply to Hume, and the reply is far more compelling now than in the eighteenth century.

Many constructivists would reply that they do not deny progress in general, but only linear and cumulative progress. While it is true that history does not reveal a simple linear progression of theories, for some of the reasons Kuhn saw, it does nonetheless build on earlier achievements so that successor theories must be at least as powerful as their predecessors to win the hearts of the new generation of scientists. When theory B replaces theory A, B explains all that A explained and more besides. Later theories may assimilate earlier theories as Einstein's did Newton's.[35] In the process, they may transform and apply limiting conditions to the earlier theories, and present terminology and theorems that are incommensurable with those of the previous theories. But they do not simply dismiss or overthrow successful theories of the past without improving upon them, and in that sense transcending them.

The preceding argument applies to the relativist, constructivist attacks on science. Instrumentalists, however, do not deny the fact of scientific success, only that the theories used in science represent the structure of reality. Their approach is a kind of skepticism about the theoretical posits of science, of quarks, electrons, even molecules. They can accept scientific theories as providing valuable predictions and correlations, but not as giving deep explanations of the world. In some sense I do not need to argue against them insofar as I present a case for scientific progress. Yet at another level I want to question their skepticism. According to instrumentalism, the alleged explanations of science do not convince. Successful scientific theories succeed in predicting, but since we cannot compare the unobservables to reality, we cannot verify the truth of their claims; they remain instruments of action rather than representations of reality. On this account much of what science is currently doing remains inexplicable. Having determined the chemical and nuclear properties of atoms, science would seem to have reached the limits of what might be asked in the way of prediction. Yet now scientists are testing theories about the structure of atoms—at present the search for the elusive Higgs boson is bearing fruit. In fact, the only reason to hunt for things like this must be to confirm a certain theory about the structure of matter.

---

35. I recognize this point is controversial, but I do not think it should be. I recently heard a professor of physics report that he had once tried to teach an introductory class relativity theory without first teaching Newtonian mechanics. The educational experience was a disaster for the students: they could not understand Einstein's theory without first understanding Newton's. Far from seeing the practitioners of different paradigms as living in different worlds, we may need to see them as functioning at different stages of reflection.

In essence, having determined the mass, charge, spin, and lifespan of subatomic particles, some (like mass) to many decimal places, we are looking for other entities and properties that our theory tells us *should* exist.[36]

One can get the impression from reading in certain fields that the whole approach of scientific realism is passé, leaving only the question of what form of anti-realism should replace it; as one philosopher put it bluntly at the beginning of an article, "Realism is dead."[37] Yet reports of the demise of scientific realism have been greatly exaggerated. The arguments for it are stronger than ever, and while many details of a complete theory remain to be worked out, the outlines of a satisfactory theory are available.[38] There are many technical issues involved in a complete argument for scientific realism that belong properly in a work dedicated to the philosophical problem rather than a historical study such as this one. Yet one dimension of an adequate defense of such a view is its ability to find evidence for its claims in the history of science.

1.7.3 History of Science Without History

Is there objective historical knowledge? The obvious argument for the objectivity of scientific knowledge is the increasing ability of science and science-based technology to predict and manipulate the world. Scientific knowledge involves a high level of consensus about

---

36. I recognize that instrumental is a more subtle position than many and needs more attention than I can give it here. My remarks here can only be general and programmatic. There is a strong historiographical tradition according to which Greek astronomy was carried out in an instrumentalist way, as a project to "save the phenomena," apparently originating with Plato. See Duhem 1969 [1908]; Natorp 1921; Wasserstein 1962; for discussion and criticism, see Mittelstrass 1962; Vlastos 1975, ch. 2; Lloyd 1978; Mourelatos 1981; Hetherington 1996.
37. Fine 1984:83. And later, "realism is well and truly dead, and we have work to get on with, in identifying a suitable successor" (84).
38. See Newton-Smith 1981; Boyd 1984; Leplin 1984; Miller 1987; Norris 1997; Chakravartty 2007; Devitt 2010, inter alios.

basic truths, even while there is often great disagreement on questions in areas of new research. In the realm of history, Aviezer Tucker has argued for an analogous kind of objectivity: "consensus on historiographic beliefs in uncoerced, heterogeneous, and sufficiently large groups of historians is indicative of knowledge of history."[39] That is, despite different starting points, theories, and assumptions, historians do achieve agreement over large domains, like scientists. Their researches are constrained by documentary and other evidence that can be disputed in detail but that nevertheless elicits overall agreement. It seems incontestable that we understand the past better now than we did fifty years ago, and much better than one hundred years ago, and so on. While there are inevitably many areas of disagreement in historiography, this should not blind us to the many more areas of agreement and the growing body of evidence and consensus.

The fact that history is knowledge of the particular, and often moves from particular premises to particular conclusions (even if it sometimes uses generalizations), should not deter us from calling it knowledge. After all, important areas of science depend on historical types of knowledge, for instance paleontology and even modern cosmology. If the epistemological models of Plato, Aristotle, and Hempel, which require deduction from universal premises, do not allow for such knowledge, so much the worse for those models.

There are further challenges for historiography. Some claim that the historian should so completely immerse himself in the past as to ignore present outcomes of past ideas:[40]

> The *diachronical* ideal is to study the science of the past in the light of the situation and views that actually existed in the past;

---

39. Tucker 2004:23. See also Windschuttle 2000.
40. Although Skinner 1969 makes good criticisms of some problematic approaches in intellectual history, he seems to fall into this position.

in other words, to disregard all later occurrences that could not have any influence on the period in question.... So, ideally, in the diachronical perspective one imagines oneself to be an observer *in* the past, not just *of* the past.[41]

The virtue of this approach is to try to purge oneself of contemporary prejudices. The problem is that the approach does not allow one to find any relevance between past and present; it is a purely antiquarian approach. The alternative is the *anachronical* approach that sees from a contemporary perspective. Helge Kragh concludes that "in practice the historian is not confronted with a choice between a diachronical *or* an anachronical perspective. Usually both elements should be present...."[42] Perhaps it is best to think of a diachronical moment and an anachronical moment in research or exposition: the former allows one to see the problems faced in the historical situation, the latter to appreciate the outcomes of the actions taken and their relevance for a present understanding. In the present book, I will insist on a study of the historical situation for seeing why certain moves were important innovations in their time; and on the relevance of subsequent developments for appreciating the importance of those innovations.

Connected with the diachronical approach is a theory that demands historical events and actions be understood in terms of the agents' consciousness.[43] This theory sometimes appears in the guise

---

41. Kragh 1987:90, italics in original.
42. Kragh 1987:107, italics in original. He also observes, "The history of science is not a two-part relationship between the historian and the past, but a three-part relationship between the past, the historian and a present-day public" (105), to which I would just add a fourth relatum, present-day science.
43. In a classic study, Collingwood 1946 says "Historical knowledge is the knowledge of what the mind has done in the past, and at the same time it is a redoing of this, the perpetuation of past acts in the present.... To the historian, the activities ... are not spectacles to be watched, but experiences to be lived through his own mind; they are objective, or known to him, only because they are also subjective, or activities of his own." This expresses an idealist approach to history.

of studies of concepts used in the history of philosophy, science, or ideas. The rise of philosophy, for instance, may be studied through the use of the word "philosophy."[44] Such a study can be important for appreciating the professionalization of a field. Yet I take it that the fact of having and using a word for "philosophy," "science," or "astronomy," for instance, is neither necessary nor sufficient for philosophizing or conducting scientific or astronomical inquiry. The correspondence between words and activities (for instance) is never perfect, and vocabulary often lags behind practice. As Tucker says, "understanding historical agents in their own terms, ignoring the vast evidence for the factors that change their consciousnesses, can offer only a shallow and superficial knowledge of history" (202). Accordingly, in this study I do not worry a great deal about the vocabulary of science or astronomy, though I will use some ancient labels for phenomena (as well as some modern ones), and I would welcome further research on the developing vocabulary of science. (This is not to say that philology is not important for understanding the texts, only that the attribution of views to ancient writers is not a straightforward matter of their having a set vocabulary.) Thus the fact that the English word "scientist" was not coined until the nineteenth century does not make me hesitate to call Isaac Newton a scientist; nor would finding out that Anaxagoras called himself a *sophos* (if he did) make me hesitate to say he was conducting both philosophical and scientific investigations. Here I think an anachronical perspective works perfectly well.

1.7.4 *History of Science Without Science*

As to the new historiography of ancient science with its implications of History of Science Without Science, it looks suspiciously like an attempt to turn a relativist and constructivist criticism of science into an

---

44. See, e.g., Laks 2006.

ideology (see § 1.7.2 above).[45] To the extent that these isms are problematic, so will be their historiographical offspring. The proponents of the new approach can rightly point out practical shortcomings of the old historiography: it is too culturally biased, too narrowly focused on certain practices, too simplistic in its consideration of contributing factors. But if they go on to claim that the right approach is to study all cultures equally, all practices indiscriminately, all factors non-objectively (whatever that means), it seems to me that what they are in effect advocating is not a more comprehensive history of science but a wholesale sell-out to cultural studies. One essayist among the new historians of ancient science, Geoffrey Lloyd, who avoids the extreme positions of others, distinguishes between observation and theory, on the one hand, and actors' and observers' categories, on the other (1992:564). He rightly points out that "the history of science is inevitably evaluative: it always presupposes a conceptual framework and a methodology. Our best hope, as historians, is to be as self-conscious and as self-critical as we can about these—not that we shall thereby escape being evaluative" (566). We cannot, accordingly, escape using *our* theories and methods and *our* categories in studying, and evaluating, our predecessors.[46] Without *some* normative conception of science, I would add, there can be no history of science. Let me hasten to point out that many of the new historians of ancient science seem to me to be much better historians

---

45. "Much of my argument has been based on the anthropological perception that science is not the apprehension of an external set of truths that mankind is progressively acquiring a greater knowledge of, but that rather the sciences are products of human culture" (Pingree 1992:563). This "perception," however, is not an argument but a question-begging assumption, or perhaps more charitably, just a methodological presupposition for work in the field. Its justification as a principle will depend on philosophical arguments—unless anthropology turns out to be an external set of timeless truths.

46. I take it that the Lloyd's distinctions are roughly equivalent to those of the diachronical vs. anachronical perspectives in Kragh (see above, ch. 1.5). I should note that Rochberg (1992) sees the need to distinguish between truth and falsehood (549), and is hopeful that tradition "transcends culture and language" (550). But how it could under the constraints proposed by other essayists I do not know.

than their theorizing about history might suggest.[47] Indeed, they have helped to rewrite the history of early science—not, however, by admitting "imaginary phenomena" or reinventing science, but by showing how, for instance, the pursuit of astrological divination produced data and theories that had objective scientific value. Lest my remarks seem too dismissive and my own assumptions too vague or opaque, I have provided some further arguments, and offer my own minimal conditions for science and its historical study below (see Appendix 2).

1.7.5 Objective History of Science

What in past generations could have been an easy introduction to the topic of ancient science has become something like a medieval *disputatio*, with a long list of objections and replies before we reach the answer, or perhaps an Aristotelian engagement with the *endoxa*, or authoritative views, before the argument proper. Unfortunately, very little can be taken for granted in the present state of research. While my own replies should be understood as programmatic rather than demonstrative, I have attempted to point to more substantive contributions to the methodological debates. The upshot of my preliminary argument is that objective knowledge is possible both in historiography and in science. Although science is a human enterprise that has a history grounded in a cultural setting, it does not follow that scientific knowledge is merely a product of human conventions, or that the history of science is a bag of tricks we play on the dead.

All the challenges to objectivity in history and science we have met seem to come back to a few basic claims about the relativity of knowledge. If I cannot single-handedly refute all these challenges, I

---

47. See, e.g., Lloyd 1992 in this collection, and much of his research, e.g., Lloyd 1979. Also Rochberg 2004 offers real insights into a non-Western astronomy, and does so by making distinctions between scientific and non-scientific aspects of Mesopotamian astrology, while presenting what I should call a rigorously objective argument throughout.

can hope that the arguments presented have indicated that the case for relativity is at least as problematic as the case for objectivity. Further, the relativist faces a pragmatic objection that goes back at least to Plato,[48] according to which he cannot consistently advance an unrestricted thesis about anything, but can claim only that some statement is true for culture C or historical period H or paradigm P. Yet we *seem* to be able to grasp the significance of lessons from different cultures, historical periods, and paradigms in an objective way; and indeed the relativist must tacitly rely on our grasp of such lessons to draw his universal conclusions. At this point the lessons of history must be the touchstone for methodological claims about scientific progress or the lack thereof. (But if there is no historical knowledge, there can be no evidence about the methodology of science.) Ultimately, if science is objective knowledge, that fact should be discernible in the history of the subject itself.

This brings me at long last to my own story. It is the account of early Greek astronomy I shall pursue in the remainder of the book. It will be unapologetically "continuist" and "progressivist." It will sound a lot like old-time history of science, without, I hope, beating the drum for Greek cultural superiority, or assuming that each step forward follows simply and inevitably from the previous step. Yet what a scientific attitude offers is precisely what some critics want to avoid—the possibility of an improving match between theory and evidence. One of the virtues of natural science is its ability to correct its own conceptions, to make ever more accurate determinations and predictions. If there is progress in science, then there are discoveries and advances in its history, which can be discerned at least in retrospect, or anachronically.

If our picture of the origins of Western astronomy is fuzzy, it is partly because there are many factors and activities contributing to it; but partly because researchers have been looking for science in all the

---

48. *Theaetetus* 170d–171c.

wrong places. Some look to grand theories unconnected with their treatment of evidence (advocates of Footnotes to Thales, Pythagoras). Some seek to redefine science so that any explanation can count regardless of method or content (advocates of History of Science Without Science). Some seek to dissociate science from substantive progress altogether (advocates of Science Without Knowledge). What they all miss is what makes science scientific: its ability to get things right, and to improve successively on its own understandings. If, however, substantive progress is what characterizes science, progress itself can serve as a kind of criterion for identifying instances of science in history. The die-hard skeptic will no doubt reply that the criterion cannot be satisfied—not now, and a fortiori not in early Greek times. We shall see.

The locus classicus for decisive scientific progress is the so-called scientific revolution that began roughly with the publication of Copernicus' almost prophetically titled work *De revolutionibus* in 1543.[49] But I think there is good evidence that some episodes of scientific progress, some discoveries, occurred soon after scientific attitudes to the world emerged in Greece. I recognize that such a claim will be highly controversial, conflicting with standard assessments of the Presocratics found in historical treatments of them as philosophers and

---

49. The rapidity of the advances is relative—a century and a half from the *De revolutionibus* to Newton's *Principia*. But the changes were thoroughgoing, profound, and irreversible. Incidentally, Steven Shapin, who tends to focus on the sociology of science, begins his book *The Scientific Revolution* with the provocative statement, "There was no such thing as the Scientific Revolution, and this is a book about it" (1). I have no special commitment to the Scientific Revolution, or scientific revolutions in general (if there were no such revolutions, the history—and philosophy—of science would be considerably less problematic). But as far as I can tell from the book, Shapin gives no real argument for his stated thesis, just a valuable study of the Scientific Revolution, so that the book (and the otherwise misleading title) could have been the same or better (i.e., self-consistent) without the first sentence. Shapin seems to instantiate the new historiography of science in some ways (see, e.g., 161–62), but in other ways his historiography in the work cited is not new at all, and may be as non-revolutionary as his putative subject.

non-scientists, as well as with present-day relativistic attitudes about science in general. Nevertheless, I will argue that the Presocratics engaged in *successful* scientific research already in the early fifth century BC. Secondarily I will try to show that the advances they made had a significance that reached beyond the confines of their particular research projects to influence later science—specifically, astronomy—reaching down to the present. While the putative discoveries they made grew out of the questions they asked from their limited perspectives and the assumptions they made based on their parochial worldviews, the results they arrived at transcended their culture and their age. They established interpretations of the world that, having survived unchanged and unrefuted for almost two and half millennia, merit the name of scientific truths if any interpretations do. If this is so, there is a lesson to be learned from the Presocratics about the possibility of cumulative objective knowledge in science. The historical thesis concerning early Greek astronomy I will pursue will support a philosophical thesis about the possibility of objective knowledge in general and scientific knowledge in particular.

If the story I am about to tell is true, it can offer just one little counterexample to the claims that there is no scientific progress and that science never arrives at the truth. But that is the beauty of counterexamples: it only takes a homely little one to bring down a big, glamorous theory.

Chapter 2

Azure Pastures

An Early Ionian Model

The sun's flame grazes in the azure pastures of the sky.
—Lucretius, *On the Nature of Things* 1.1090

To tell the story of early Greek astronomy it is necessary to begin with the beginnings of speculation. Before there was philosophical speculation on the topic, there was a mythological tradition of theogony. The origins of this tradition are shrouded in the mists of prehistory, but we can pick up the latest stage of it, in the works of the poet Hesiod.

2.1 HESIOD'S MYTHICAL COSMOGRAPHY

A rude shepherd from the Boeotian town of Ascra, Hesiod credits his poetic gift to an encounter with the Muses on the slopes of Mt. Helicon. They "breathed into [him] divine voice," gave him a laurel staff, and taught him about the origins of the world, the gods, and men.[1] His epic poem, the *Theogony*, tells the story of the creation, or rather, the birth of heaven and earth. Although the story is uniquely Hesiodic, the general outlines of his tale are part of a common cultural

---

1. *Theogony* 22–34.

heritage, as can be seen by comparison with statements in Homer and other mythological sources.

After an invocation of the Muses, he gives this account of the origins of the world:

> Indeed, first was Chaos born, but then
> broad-bosomed Earth, steadfast seat ever of all
> [the immortals, who hold the peaks of snowy Olympus],
> and misty Tartarus in a recess of the wide-wayed earth,
> and Eros, who fairest among the mortal gods,
> looser of limbs, of all gods and all men
> overcomes the thought in their breast and their wise counsel.
> From Chaos Erebus and black Night were born,
> And from Night Aether and Day were born,
> whom she bore being with child after mingling in love with Erebus.
> And Earth first bore equal to herself
> starry Heaven, that he might cover her all around,
> that he might be a secure seat for the blessed gods always.
> And she bore long Hills, lovely haunts of the gods,
> and Nymphs, who dwell on the woody hills.
> And she bore the fruitless deep, with raging swell
> Sea, without desirable love. But then
> lying with Heaven she bore deep-swirling Ocean,
> Coeus, Crius, Hyperion, Iapetus,
> Theia, Rhea, Themis, Mnemosyne,
> golden-crowned Phoebe and lovely Tethys.
> After them was born the youngest, wily Cronus,
> most terrible of her children. And he hated his flourishing father.
>   (104–38)

What we have here is not the creation of the world by an all-powerful God, such as was taught in the Judaism of the same time, but a series

Chapter 2

Azure Pastures

An Early Ionian Model

> The sun's flame grazes in the azure pastures of the sky.
> —Lucretius, *On the Nature of Things* 1.1090

To tell the story of early Greek astronomy it is necessary to begin with the beginnings of speculation. Before there was philosophical speculation on the topic, there was a mythological tradition of theogony. The origins of this tradition are shrouded in the mists of prehistory, but we can pick up the latest stage of it, in the works of the poet Hesiod.

2.1 HESIOD'S MYTHICAL COSMOGRAPHY

A rude shepherd from the Boeotian town of Ascra, Hesiod credits his poetic gift to an encounter with the Muses on the slopes of Mt. Helicon. They "breathed into [him] divine voice," gave him a laurel staff, and taught him about the origins of the world, the gods, and men.[1] His epic poem, the *Theogony*, tells the story of the creation, or rather, the birth of heaven and earth. Although the story is uniquely Hesiodic, the general outlines of his tale are part of a common cultural

---

1. *Theogony* 22–34.

heritage, as can be seen by comparison with statements in Homer and other mythological sources.

After an invocation of the Muses, he gives this account of the origins of the world:

> Indeed, first was Chaos born, but then
> broad-bosomed Earth, steadfast seat ever of all
> [the immortals, who hold the peaks of snowy Olympus],
> and misty Tartarus in a recess of the wide-wayed earth,
> and Eros, who fairest among the mortal gods,
> looser of limbs, of all gods and all men
> overcomes the thought in their breast and their wise counsel.
> From Chaos Erebus and black Night were born,
> And from Night Aether and Day were born,
> whom she bore being with child after mingling in love with Erebus.
> And Earth first bore equal to herself
> starry Heaven, that he might cover her all around,
> that he might be a secure seat for the blessed gods always.
> And she bore long Hills, lovely haunts of the gods,
> and Nymphs, who dwell on the woody hills.
> And she bore the fruitless deep, with raging swell
> Sea, without desirable love. But then
> lying with Heaven she bore deep-swirling Ocean,
> Coeus, Crius, Hyperion, Iapetus,
> Theia, Rhea, Themis, Mnemosyne,
> golden-crowned Phoebe and lovely Tethys.
> After them was born the youngest, wily Cronus,
> most terrible of her children. And he hated his flourishing father.
> (104–38)

What we have here is not the creation of the world by an all-powerful God, such as was taught in the Judaism of the same time, but a series of births of divine figures, many of whom have cosmological roles and characters. "Chaos" signifies not disorder, but a yawning gap, a great space that opens up like a kind of primordial womb. (The word "chaos" is neuter, however, leaving us with an uncertain gender for the first, mysterious being.) From Chaos is born Gaia, broad-bosomed mother earth, who will bear many other children. Then Tartarus, the underworld, arises, then Eros, god of love, who will serve as agent for sexual unions giving rise to future births.

There is some kind of logical connection between parents and offspring. The chasm Chaos gives birth to Erebus (darkness) and Night; Night gives birth to Aether (the shining upper air) and Day. Thus sometimes like gives birth to like, sometimes opposite gives birth to opposite. Heaven and Earth produce Ocean, which is the outer sea surrounding the disk-shaped earth, where the two parents meet. Earth produces Sea (Pontus), the inner waters, without a partner. Earth produces Hills, located on her, and Nymphs, dwelling on the Hills. So one feature sometimes begets another feature contiguous with it. But the diversity of connections shows that there is no single law of inheritance covering all cases.

At the end, wily Cronus is born, who will displace Heaven as the leader. He castrates his oppressive father with a sickle. He begets children with Rhea, but swallows them until his son Zeus, who was hidden by his mother, matures and challenges him for rule. A war ensues between the Titans, led by Cronus, and the Olympians, led by Zeus, which the Olympians win to take control of the world. The losers are confined in Tartarus, the gloomy underworld, which becomes their dungeon. Thus there is a mixture of cosmography, war, and politics in the story of the origin of things. Some of the bizarre stories of conflict among the gods are paralleled in other myths from neighboring Indo-European peoples, and could be inherited from early mythology.[2] While some scholars have argued for extensive Greek borrowings from Asian cultures, both before and after Thales, what can be proved concerning intellectual borrowings is small.[3] Although Greek science makes its own contributions independently of whether its key ideas come from the Near or Middle East, it remains remarkably difficult to find any theoretical borrowings from Asia, as we shall have occasion to see hereafter.

The picture of the world seems to be a notion shared by early Greek mythographers. Homer depicts a circular earth surrounded by Ocean, with a heaven above and an underworld beneath.[4] According to Hesiod,

> as far below earth as heaven is above
> ... so far is misty Tartarus from earth.
> For a bronze anvil falling for nine days and nights
> from heaven would reach earth on the tenth;
> and again a bronze anvil falling nine nights and days
> from earth would reach Tartarus on the tenth.
> Around it runs a bronze fence. And around it night
> is poured out in triple rows like a necklace; but above
> grow the roots of earth and the fruitless sea. (*Theogony* 720–8)

By this quaint comparison, Hesiod measures the cosmos: an anvil falling from heaven would take ten days to reach the earth, and one

---

2. In particular the Hittite Kumarbi epic shows a succession of gods with stories similar to those of Hesiod. See Ancient Near Eastern Texts 1969:120–21. The Hittites are Indo-European, at least linguistically. Common myths in this case are compatible with a common cultural background rather than contemporary borrowing, especially since the Hittites did not have any close contact with the Greeks, as far as we know. Scholars often treat the Hittite epic as Near Eastern rather than Indo-European in content, but without much argument. It may, of course, reflect elements of both traditions.
3. For intellectual borrowings, see West 1971; West 1994; West 1997. For general mythological background, see Frankfort et al. 1949. For a survey of likely contacts, see Burkert 1992.
4. *Iliad* 18.607–8, cf. 489; *Odyssey* 10.508–14; 11.13–15.

falling from the earth would take ten days to reach the bottom of Tartarus. Thus the cosmos has a radius of ten anvil-days, and is perhaps roughly symmetrical on both sides of the earth. The earth is flat and circular in shape and ringed by a body of water, Ocean. Inside the disk is an inland sea. The region above the earth is bright and airy, that below is dank and gloomy. Various gods inhabit the several realms and rule over portions of them.

Thus, we find in Hesiod a mythical cosmography. A genealogical process produces the world as we know it, which consists of cosmic deities with their own histories and characters. After the cosmic deities come Olympian deities who succeed to power and influence, and come to rule over regions of the world. Hesiod provides a point of departure for subsequent philosophical cosmologies, which exhibit both continuities and differences.[5]

2.2 IONIAN THEORIES

Suddenly, a new style of cosmology arose in the sixth century. Rather than divine beings, the heavenly bodies and features of the earth were seen as physical objects. They were things rather than persons, objects rather than subjects.[6] Invariant physical regularities were to be studied rather than arbitrary acts of agents. The Ionian philosophers produced a bewildering number of distinct theories of the world. G. E. R. Lloyd observes, "There is no such thing as *the* cosmological model, *the* cosmological theory, of the Greeks. One is hard put to describe the *predominant* notion of notions in Greek cosmology...."[7] Since there are no generally

---

5. For a study of Hesiod in relation to his successors' cosmologies, see Stokes 1962, 1963.
6. This is not to say that there is no continuity or connection between mythological and philosophical models. On this, see especially Stokes 1962, 1963.
7. Lloyd 1975:205, Lloyd's italics. His topic goes beyond the Presocratic period, but includes it prominently. He goes on to describe three general traits of Greek cosmology, but he does not identify periods or a direction of development.

recognized groupings of Presocratic cosmologies, there are ipso facto no historical groupings, and hence no standard accounts of historical development, much less of progress, and more generally no record of increasing adequacy, either theoretical or empirical.[8] To the contrary, I believe that there are important historical differences that emerge in the course of early cosmological theorizing.[9] These can become apparent as we look at the characteristics of sixth-century BC theories in relation to fifth-century theories. My hope is to reveal a progression in theories and to show how this progression embodies a theoretical development and, surprisingly, an empirical foundation. The place to start is with the earliest philosophical theories, those of the sixth century, and to review them with an eye to features they have in common.

2.2.1 Thales

The first philosopher or cosmologist was, according to Aristotle and most early authorities, Thales. He can be dated by his success in "predicting" an eclipse, most likely that of 585 BC.[10] Soon afterwards he was

---

8. The few attempts to see a clear philosophical and/or scientific development include David Furley 1987, who sees a development of a contrast between the Infinite Universe and the Closed World (the distinction from Alexander Koyré), associated with a contrast between "linear" or "parallel" dynamics, whereby bodies fall straight down, and "centrifocal" dynamics in which heavy bodies move toward the center of the universe (see Furley 1989: 14–26). At about the same time Detlev Fehling 1985 argued for a gradual development of cosmological theories based on conceptual advances; his view has some important similarities to the view I will develop here, but it entails a wholesale rejection of much of the doxographical record and seems too radical to be plausible (see also Fehling 1994). Recently, Sedley 2007 has argued for a contrast between mechanistic and creationist cosmologies. I will not follow this approach, though it produces valuable insights. Gregory 2007 provides a recent study of cosmogonies, without noting major patterns of development for the Presocratic period. Some studies of cosmology that are topically rather than historically organized, e.g. Sambursky 1956, Wright 1995, tend to preclude considerations of historical development.
9. I looked at this from the standpoint of scientific or proto-scientific methodology in Graham 2006, but in the process of that study I came to see that there was much more to be said about actual scientific practice and knowledge.
10. Stephenson and Fatoohi 1997.

recognized as one of the seven sages, possibly because of this feat.[11] He seems not to have left a book, and hence is known only through hearsay and tradition.[12] We are interested in his cosmological and astronomical theories, about which there is limited information. Yet this information provides a fair number of claims about Thales' advances in astronomy and cosmology. It will be helpful to provide a list of these, making a few preliminary connections in a doxographic style:

1. Thales founded the study of astronomy among the Greeks.
2. The earth floats on water like a log or raft.
3. The stars are carried around by water.
4. The stars are fueled by exhalations or vapors of water.
5. The stars are earthy with fire in them.
6. The year is 365 days long.
7. Thales described the path of the sun from solstice to solstice, predicting solstices and discovering that the seasons were unequal in length.
8. The Pleiades set on the 25th day after the equinox.
9. Thales called the last day of the month the 30th.
10. The moon is illuminated by the sun.
11. Thales predicted a solar eclipse.
12. Solar eclipses are caused by the blocking of the sun's light to the earth.
13. Lunar eclipses are caused by the blocking of the sun's light by the earth.
14. The ratio of the sun's size to its orbit is 1:720 (or 750).
15. The stars of Ursa Minor form the constellation closest to the north pole.

---

11. Diogenes Laertius 1.22, on the authority of Demetrius of Phaleron; see O'Grady 2002:146, 273–76; Rossetti 2011.
12. See Kirk, Raven, and Schofield 1983:86–88. A *Nautical Star Guide* is attributed to him by some sources; it is also attributed to Phocus of Samos (Diogenes Laertius 1.23).

Let us look at these claims individually.

1. *Thales founded the study of astronomy among the Greeks.*[13] This judgment goes back to Eudemus, a student of Aristotle, who wrote a Study of Astronomy identifying stages in the development of astronomical thought.[14] Eudemus follows Aristotle's tentative judgment that Thales was the first natural philosopher.[15] It is likely to be true in the sense that Thales may have been the first thinker to look at the stars in a scientific way, to see them as physical objects with natural motions rather than autonomous divinities. It is, of course, not true in the sense that he was the first Greek to track the stars and try to understand their seasonal occurrences. Hesiod, for instance, uses the stars as guides for planting and other practical activities. That Thales put astronomical theory on a scientific footing by, for instance, developing a complex theory of the heavens is questionable (see on [3] and [15]).

2. *The earth floats on water.*[16] This opinion is known to Aristotle; he reports it as hearsay, as he typically does of all Thales' doctrines, showing that he has no primary source materials. Yet at least we can infer that it belongs to the earliest collection of his views, presumably found in Hippias' *Collection*.[17] We are not told the extent of the sea on which the earth rests, but it may be infinite, or at least indefinitely extended. This would preclude the possibility of heavenly bodies moving under the earth. It would imply that the earth is more or less

---

13. Diogenes Laertius 1.23 = DK 11A1; Hippolytus *Refutation of All Heresies* 1.1.4. For a new edition of Thales with greatly expanded testimonies, see Wöhrle 2009.
14. The title of Eudemus' work, *Astrologikē historia* does not necessarily mean "history of astronomy"; rather *historia* generally means "research" and only in special contexts "history." See Mejer 2002b. No doubt the work contained some sort of historical survey or reflections as part of the study of astronomy. For the work as historical, see Bowen 2002, who takes the work as doxographical and therefore historical; but doxography is typically dialectical rather than chronological in organization, as Mansfeld has shown in multiple studies, and in any case history is much more than chronology.
15. *Metaphysics* 984a2–3.
16. Aristotle *On the Heavens* 294a28–33 = A14; Seneca *Natural Questions* 3.14.1.
17. See Patzer 1986:33–42; Snell 1944; Classen 1965; Mansfeld 1990:126–46.

flat and of small size relative to a vast or limitless surface of water on which it floats.[18] This seems to be a reliable report if any is.

3. *The stars are carried around by water.*[19] This report is late, found in Hippolytus, and vague; it may indicate only an attempt to bring Thales' cosmology into the fold of natural philosophy as it later developed. The doxographers (ancient collectors of philosophic opinions; see below, ch. 6.5) ascribed some sort of motive power to water to account for the motion of the heavenly bodies.

4. *The stars are fueled by exhalations of water.*[20] On this account, the heavenly bodies are fueled by moist vapors arising from the sea, in the manner of a lamp that burns olive oil. This kind of theory seems consistent with early conceptions of astrophysics, as we shall see. But it also seems to arise from an attempt to infer the phenomena from the simply theoretical assumption that all is water. From Aristotle on, commentators tried to fill in the blanks of sketchy theory.[21] Presumably Theophrastus, expanding on Aristotle, made this connection speculatively and thus brought Thales into the Ionian tradition.

5. *The stars are earthy with fire in them.*[22] This report appears late in Aetius. It has no pedigree to connect it with the early reports of Thales, and it may constitute an attempt to supply missing information. See comments on (item 10).

6. *The year is 365 days long.*[23] This is a late attribution, appearing in Diogenes Laertius. It could, however, represent a true tenet of Thales, since the year of 365 days originated in Egypt and Thales has other reasonably well-attested ties to Egypt.[24] Thales would not have

---

18. O'Grady 2002:87–94 disputes the claim that the Earth floats on water, claiming that Thales held rather that portions of earth floated on water as though they were islands. She also claims that Thales advocated a spherical earth, following Aetius 3.10.1. Against both of these views, see Couprie 2011:64–67.
19. Hippolytus *Ref.* 1.1.2–3.
20. Aetius 1.3.1.
21. *Metaphysics* 983b25–27; Simplicius *Physics* 23.21–29.
22. Aetius 2.13.1 = A17a.
23. Diogenes Laertius 1.27.
24. Neugebauer 1957:81.

needed to discover this year, merely to accept the authority of the Egyptians on what was the correct length of the year to the nearest whole number of days.

7. *Thales described the path of the sun from solstice to solstice, predicting solstices and discovering that the seasons were unequal in length.*[25] The chain of evidence in this case goes back to Eudemus, a fairly good source. But the claims are problematic. It is difficult to determine the solstices from observation of the northernmost rising (in summer; southernmost in winter) of the sun because for several days around the solstice the sun hardly seems to move. (In fact it is easier to track the year from equinox to equinox with the use of a gnomon.) We have other evidence that it was Euctemon who actually determined the different lengths of the (astronomical) seasons around 430 BC marking "a comparatively sophisticated stage in astronomical thought."[26] But even Euctemon's system may have been schematic and derivative, not based on his own observations.[27] Now there is one way one might "predict" a solstice without advanced methods: one might do one's best to identify one solstice and then use the 365-day year to anticipate the next. But in general this alleged feat of Thales seems unlikely.[28]

8. *The Pleiades set on the 25th day after the equinox.*[29] The risings and settings of stars was part of the almanac-like lore that Hesiod and later writers of *parapēgmata* recorded.[30] Thus there is nothing anachronistic about this notice. On the other hand, it seems to originate from *The Nautical Star Guide*, one of the probably spurious works attributed to Thales.[31] It is possible, however, that this work incorporated observations from Thales.

---

25. Diogenes Laertius 1.23, 27; Dercyllides from Theon of Smyrna 198.14–18.
26. Dicks 1970:88.
27. See Bowen and Goldstein 1988 and see below, pp. 236–37.
28. See O'Grady 2002:147–50.
29. Pliny *Natural History* 18.213 = A18.
30. See Bickerman 1980:58; Lehoux 2007. A *parapēgma* was a calendar-cum-pegboard in which one could move a wooden peg from one day to the next through the year.
31. See n. 12 above.

9. *Thales called the last day of the month the 30th.*[32] The lunar month has either 29 or 30 days in it;[33] Thales is said to have first called the last day of this cycle the thirtieth. This point is closely connected with his alleged explanation of solar eclipses in the tradition. But it is possible that he might have provided terminology solely in the interest of regulating a lunar-based calendar.[34] It is also possible that this statement marks an effort to adopt the Egyptians' thirty-day month as a calendaric standard. In any case it is difficult to evaluate this claim without more evidence.

10. *The moon is illuminated by the sun.*[35] This point is an important preliminary to a correct account of solar eclipses. If it is true that Thales recognized this fact, he would indeed be a gifted and far-sighted astronomer, inasmuch as he recognized the source of lunar light. No supporting evidence, however, is given for this claim other than his knowledge of the solar eclipse, which presupposes this point. So the reliability of this claim depends on the evidence for his knowledge of eclipses. The moon is also said to be earthy.[36] This seems closely related to the fact that the moon is opaque to the sun's light. It may provide the basis for the claim (5) that the stars (heavenly bodies) are earthy.

11. *Thales predicted a solar eclipse.*[37] This is the most remarkable accomplishment attributed to Thales. The report has a good pedigree going back to Herodotus, and, according to Diogenes Laertius, to Xenophanes and Heraclitus before Herodotus and Democritus after

---

32. Diogenes Laertius 1.24; Oxyrynchus Papyri vol. 53, no. 3710, col. ii, fr. (c), lines 36–43.
33. A synodic month, or mean period from new moon to new moon, is, by modern calculation, 29.53 days. Since a calendar tracks only full days, the cycle takes either 29 or 30 days. The Greeks would measure from the first appearance of the crescent moon in the west to the first re-appearance in its next orbit. A 30-day month was called "full" and a 29-day month "hollow" (Geminus *Introduction* 8).
34. See Bowen and Goldstein 1994 for a good exposition of differences between the Greek and Egyptian calendars in terms of identifying the first day of the month.
35. Aetius 2.28.5.
36. Aetius 2.25.8 (cf. 2.24.1 ps.Plutarch).
37. Herodotus 1.74.2; Pliny *Natural History* 2.53; Diogenes Laertius 1.23.

him. But how could he have managed such a feat? Scholars have often pointed to Babylonian methods of studying the heavens, backed up by centuries of observations. Patterns of eclipses could be discerned in the records and repetitions of cycles anticipated.[38] The Babylonians discovered a cycle of 223 lunar months (eighteen years, the "Saros" cycle) and a more accurate one of three times that length (the "Exeligmos").[39]

Yet there are problems with this method. In the first place, it could at best work for the location where the observations were made. Since solar eclipses are visible only over a narrow band of earth, the cycles of the "great year" that align the nodes of the sun and moon could ensure a repetition of the events only in a fixed locality.[40] But the centuries of observations were recorded in Babylon and Nineveh, far distant from Miletus, and would be useless for the Ionian city. (There is no evidence that Greeks ever kept the kinds of archives the Babylonians did, and in any case literacy was a recent phenomenon in the Greek world of Thales' time.)[41] At a later time, the Babylonians could predict that a solar eclipse was possible when there was a conjunction of the sun and the moon; but this technique was not available in Thales' time.[42]

Herodotus records that "This change of day [solar eclipse] Thales of Miletus predicted to the Ionians, setting as a limit that year in

---

38. Sambursky 1956:11–12.
39. Hartner 1969:67 claims he used the latter; cf. Blanche 1968.
40. Even within the locale of the observations, the method is not accurate for solar eclipses: there is a displacement of almost a third of a day, which puts the consecutive solar eclipse 115° to the west of the original one (O'Grady 2002:131).
41. "[I]t is highly unlikely that a record of dated eclipses had been kept since the eighth century in a city where even the list of eponymous *stephanēphoroi* did not begin until 525" (Mosshammer 1981:147; cf. Bickerman 1980:67).
42. The "Saros" cycle of 223 synodic months (about 18 years) seems to have been developed before 525 BC and refined to a 19-year cycle by 475 BC. This scheme could predict about 38 eclipses during the cycle, based on repetitions of earlier cycles. See Hunger and Pingree 1999:183–84, Rochberg 2004:139.

which the change actually took place" (1.74.2). There is a suggestion that he announced this prediction at an Ionian festival and promised the event within the coming year. O'Grady argues that he used a correlation that allows for a solar eclipse 23½ months after a lunar eclipse (2002:140, 142), in effect an empirical rule-of-thumb arrived at inductively.[43] This would allow him to predict an eclipse without any extensive data-base—or complex theory. It is an interesting suggestion and could account for the prediction without presupposing extensive research or advanced theory, which do not seem to have been available. It remains, however, a speculation for which there is no additional evidence than the prediction itself.[44]

The crucial point here is that there is no known way that Thales might have reliably predicted a solar eclipse in the sixth century BC.[45] He may have made a lucky prediction that was confirmed, but this was not in our terms a scientific prediction. Furthermore, whatever method he used was not handed down to posterity, making it for that reason dubious that he had any method in the first place.[46]

12. *Solar eclipses are caused by the blocking of the sun's light to the earth.*[47] According to Eudemus, Thales was the first to study solar

---

43. Similarly van der Waerden 1954:87; Hartner 1969; Couprie 2011:55–62.
44. Bowen 2002:313–14 is skeptical about what Thales actually predicted according to Herodotus.
45. "[W]hat follows from Neugebauer's research is only that, *if* Thales predicted the eclipse, he did not do so on a scientific basis" (Kahn 1970:115). "[A]t no stage was anyone, whether Greek or Babylonian, in the ancient world in a position to predict solar eclipses *visible at a particular point on the earth's surface*. . . . Moreover *if* [Thales] had access to Babylonian eclipse data it is very strange that no other Greek astronomer made any use of them down to the fourth century" (Lloyd 1991c:293-94, Lloyd's italics).
46. According to Hartner 1969:69–70, "he had learned [because the eclipse he predicted came too soon] that things are more complicated than he had believed. It was probably for this reason that he never spoke to anybody, even his closest friends and disciples, of his method and its failure. . . ." But "Hartner's ingenious approach only serves to demonstrate how utterly fictional the story of Thales' prediction is" (Mosshammer 1981:147). Mosshammer takes an extreme skeptical view that Thales made some remark about the eclipse that was later confused with a prediction.
47. Pliny *Natural History* 2.53; Aetius 2.24.1; Theon of Smyrna 198.14–18.

eclipses. The reports we have from Eudemus, however, do not go so far as to say Thales understood solar eclipses correctly. We get the following from Aristarchus of Samos, the great Hellenistic astronomer: "Thales said the sun is eclipsed when the moon moves in front of it, inferring this fact from the day on which the eclipse occurs, which some call the thirtieth, others the [day of the] new moon."[48] This recently published fragment seems to vindicate Thales as a groundbreaking theorist.[49] It hints (though it does not say clearly) that Thales' recognition of the thirtieth of the month was connected to his knowledge that solar eclipses occur only at the time of the new moon, that is, when the moon is in conjunction with the sun and can therefore block its light.[50]

Yet this is problematic. Why does no one before Aristarchus (or Eudemus, if he holds this view) know about Thales' knowledge of eclipses? In particular, Aristotle always speaks cautiously of Thales as if all knowledge about him were hearsay. Who could have transmitted such knowledge, and in what work? No reaction to such a theory is found before the fifth century. Our best fourth-century source, Aristotle, does not know of any advanced astronomy in Thales. After that time, the possibility of contamination from other theories increases. It seems likely that what happens in the post-Aristotelian period is that commentators are reconstructing Thales' thought: He predicted an eclipse; therefore he understood the mechanisms of eclipses (which were now known by philosophers); therefore we can safely attribute to him the correct theory of eclipses.[51] Yet an understanding

---

48. Oxyrynchus Papyri vol. 53, no. 3710, col. ii, fr. (c), lines 36–43.
49. Lebedev 1990 argues that Aristarchus' sources must be pre-Peripatetic.
50. For a defense of this approach on the basis of Mesopotamian knowledge, see Aaboe 1972.
51. Eudemus may be the first one to combine claims of prediction and explanation; see Bowen 2002:311–15. Bowen warns that we should not take for granted that the claim of explanation was derived from the claim of prediction; true, but the inference still seems to best account for the appearance of the latter; see Kirk, Raven, and Schofield 1983:82; Guthrie 1962–1981, 1:49. Zhmud 2006:241 argues that Eudemus did not conflate prediction and explanation, but rather those who reported his work.

of eclipses is neither necessary nor sufficient for a prediction. This can be seen by the fact that Babylonian methods could at least provide an empirical basis for prediction, but without any accompanying physical theory of the type the Greeks developed; on the other hand, the Greeks of the fifth century on could explain eclipses perfectly well but not predict them, since qualitative explanation does not by itself translate into a mathematical model that tracks the motions of sun and moon adequately.

Aristarchus' notice marks a valuable stage in the Thales reception, one that accounts for the doxographical treatment of his theory (whether it came directly from him or not): Thales must have been the discoverer of the true account of solar eclipses because he knew how to predict them. Yet this notice does not of itself guarantee that Thales understood anything about the true nature of solar eclipses. What it shows is how astronomers projected back onto Thales their own knowledge of solar eclipses.[52]

13. *Lunar eclipses are caused by the blocking of the sun's light by the earth.*[53] John Philoponus (sixth century AD) gives a textbook account of lunar eclipses, based on an understanding of the source of lunar light (10 above). Yet nowhere is Thales credited with predicting a lunar eclipse. There is no early evidence for this claim; it may also result from a backward projection of later astronomers. If Thales understood solar eclipses, how could he not also understand lunar eclipses? And so the legend kept growing.

14. *The ratio of the sun's size to its orbit is 1:720 (or 750).*[54] This notice ascribes to Thales another advanced piece of astronomical research. The diameter of the sun is one seven-hundred and twentieth part of its orbit, or ½°. This may be true of the moon as well. Hence the

---

52. On the whole Footnotes to Thales question, see Graham 2002.
53. Aetius 2.29.6 (§7 Mansfeld & Runia); John Philoponus *Categories* 118.4–25 = Th 434 Wöhrle. Only Stobaeus preserves the Thales reference of Aetius; it is omitted in ps.Plutarch.
54. Diogenes Laertius 1.24; Apuleius *Florida* 18.32.

moon has the size to block the sun when it is at conjunction, causing a solar eclipse. The method for determining the solar diameter is identified by Cleomedes as using a water-clock to time the rise of the sun and then comparing that to the length of a day (no easy task with a water-clock).[55] The fraction seems to presuppose the division of the circle into 360°, which was not borrowed by Greeks from the Babylonians until the Hellenistic era, and to require a level of precision unknown to early Greek astronomy.[56] In fact, Archimedes maintains that Aristarchus (early third century BC) "discovered" the ratio of 1:720 for the sun, a much more plausible and authoritative attribution than that to Thales.[57] Since Archimedes belonged to the same century and the same intellectual world as Aristarchus and was a leading astronomer himself, we may suppose that the former had reliable information about the latter's accomplishments. And if a discovery not made until the third century BC could be referred back to the sixth century, we have reason to be cautious of all extravagant claims on behalf of Thales.

15. *The stars of Ursa Minor form the constellation closest to the north pole.*[58] According to this report, Thales learned from the Phoenicians that Ursa Minor is a valuable constellation for navigation because it lies

---

55. Cleomedes *De motu circulari corporum caelestium* 2.75, 136.23–138.5 using a water clock in a method he attributes to the Egyptians. Macrobius *On the Dream of Scipio* 1.20.26–30 offers a method using the hemispherical sundial, which could be the method used by Aristarchus—though Macrobius' own results are poor (Stahl 1952:253); Aristarchus is said to have invented this kind of sundial (Vitruvius 9.8.1). Macrobius also attributes his method to the Egyptians. Presumably the relevant Egyptians in both cases are Alexandrians of the Hellenistic age (cf. Heath 1913:259, 311).
56. See O'Grady 2002:150–55.
57. Archimedes *Sand Reckoner* 2.137.12–14 Mugler; against Thales as the discoverer, see Heath 1913:22–23, 312. Since Aristarchus used a figure of 2° as a starting point for his calculation of the sizes and distances of sun and moon, it appears that he conducted that inquiry as a mathematical exercise rather than an empirical study (Lloyd 1979:121 and n. 328); he may have determined the apparent size of the sun later, as Heath believes.
58. Callimachus *Iambics* fr. 191.54–44 Pfeiffer; Diogenes Laertius 1.23.

nearest to the north pole of the sky. More precisely Callimachus says he "measured the little stars of the Wain," which may imply some sort of mapping. While the chain of evidence goes back only to the Hellenistic poet Callimachus, the story does not make any outlandish claims. The Phoenicians were the great sailors of the Mediterranean. They no doubt traded with Miletus and her colonies. That Thales might have learned about ways to improve navigation from the Phoenicians as he apparently learned other practical techniques from the Egyptians seems plausible and consistent with the state of knowledge in his time.[59] To recognize Ursa Minor requires no more than basic observational astronomy, yet recognizes a constellation with practical value for the voyager. Homer knows of Ursa Major as a circumpolar constellation, but not Ursa Minor, so at some point the new constellation was recognized.[60]

In his minimalist treatment of Thales, Dicks points out that sources for Thales down to and including Aristotle draw only on second-hand reports; "It is [only] when we come to the second main group of sources, i.e. writers after 320 B.C., that Thales' stature begins to take on heroic proportions..." (1959:298–99). Dicks' analysis still seems apt. The significant but limited advances of Thales become inflated as scientific thinkers of the Hellenistic age try to vindicate his insights by assigning the knowledge of their age to him. Thales predicted an eclipse; *ergo* he understood eclipses, and understood them as Hellenistic astronomers understood them; he must have measured the diameters of sun and moon, recognized the moon as an earthy body, shining by reflection of the sun's light, and so on. But, as we shall see, no one else in the sixth century had a clue about the mechanisms of lunar light, eclipses, or relative sizes. Not that they had no theories about heavenly bodies, but they show no awareness of the sorts of considerations Thales allegedly introduced. Thus it behooves

---

59. See White 2002. On the crude state of Egyptian knowledge at the time, see Tannery 1880.
60. *Iliad* 18.485–89.

us to remain skeptical about Thales' more sophisticated "discoveries." We may provisionally accept points (1), (2), (6), (9), and (15) in suitably limited versions as Thales' contributions.

2.2.2 *Anaximander*

Anaximander did write a book; although that book is lost to us, its theories were preserved in later accounts, so that we can reconstruct the author's views in light of an unbroken chain of evidence.[61] Anaximander is now generally regarded as the father of cosmology and the founder of scientific speculation about the world.[62] According to him, the earth is a flat disk like a column drum,[63] whose diameter is three times its height. It floats in space surrounded by rings, which constitute the heavenly bodies. Farthest out is the ring of the sun, which seems to be twenty-seven times the diameter of the earth, and one earth-diameter in thickness.[64] The ring is composed of fire enclosed by air or mist. The air hides the fire, except at a circular opening where the light shines out; this opening is what we think of as the sun. The sun's ring, then, occupies its orbit but is visible only at a single place. Thus the sun is like a cartwheel, or, as we might say today, a bicycle tire; we do not see the contents of the tire except at the valve stem. Similarly, the moon's ring consists of a similar structure closer to the earth, at eighteen earth diameters. The fire is less intense and hence less bright. The opening in this ring gradually gets covered and then

---

61. Thanks to Kahn 1960 we have a good reconstruction of Anaximander's theory in the context of his time. Thus the skepticism of Dicks 1970:44–46 seems excessive ("it seems best to admit that we really know nothing for certain about Anaximander's astronomical beliefs except that he regarded the earth (shape unspecified) as stationary at the centre of the universe" 46). For a new edition of Anaximander, see Wöhrle 2012.
62. This is the thesis of Kahn 1960, which remains to this day the bible of Anaximander studies.
63. The construction of monumental columns by stacking column drums on top of each other was a new technique in Anaximander's time, recently imported from Egypt. See Hahn 2001:149–62 et passim.
64. See Diels 1897; or the radius of the sun's circle may be twenty-seven earth diameters: Hahn in Couprie, Hahn, and Naddaf 2003:145–48; Couprie 2009a.

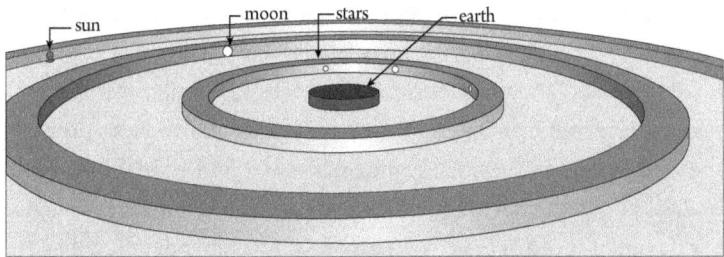

Figure 2.1. Anaximander's Cosmos. Heavenly bodies are wheels from which light shines out or fire bursts out as from a jet at an opening.

uncovered, yielding the phases of the moon.[65] Eclipses of the sun and moon occur when the opening is blocked by some process we are not told about. Finally, closest to the earth, perhaps at nine earth diameters, is the region of the stars, which presumably consists of many small rings circling around the earth (see fig. 2.1).[66] What precisely is the configuration of the stars' orbits, is not clear; they could form a sphere around the earth, or they could form concentric circles (around the north pole) without traveling under the earth.[67]

Anaximander raises the question of why the earth remains in its place, now that it is freestanding rather than rooted in place as in Hesiod[68] or floating on water as in Thales. Why does it not fall, since it is a heavy body? Anaximander answers that it stays in place by being in

---

65. Diogenes Laertius (2.1) says "the moon is a pseudo-luminous body illuminated by the sun." This is a confused report, probably caused by Diogenes' expanding an abbreviation of the name "Anaxagoras" incorrectly; see Mejer 1978a:22, 25–28. A similar attribution is made to Anaximenes; see below, n. 74.
66. Couprie has proposed using virtual cylinders for the rings of the heavenly bodies, to account for their seasonal variations: Couprie, Hahn, and Naddaf 2003:224–28; Couprie 1995. In the case of the stars this is difficult: to account for the nearness of stars to the north pole, the cylinder of stars must be prolonged, making some stars far more distant from the earth than the diameter of the cylinder. In any case, the doxography assigns plural rings to the stars: Hippolytus *Refutation* 1.6.5 = A11; Aetius 2.16.5 = A18.
67. McKirahan 2010:40 represents the stars as forming a sphere around the earth; but the textual evidence does not entail that interpretation.
68. *Theogony* 728, quoted above.

equilibrium: since it has rotational symmetry, it has no tendency to move one way or another. He gives no particular statement about the heavenly rings. They may be where they are for similar reasons; certainly their symmetrical placement relative to the earth helps preserve the earth's position.[69] Evidently the rings rotate, for we see the heavenly bodies, that is, the openings, moving across the sky each day or night. We do not hear of a definite reason for this. Anaximander describes the origin of the cosmos as resulting from a primordial movement:

> He says that that part of the everlasting which is generative of hot and cold separated off at the coming to be of the world-order and from this a sort of sphere of flame grew around the air about the earth like bark around a tree. This subsequently broke off and was closed into individual circles to form the sun, the moon and the stars. (ps.Plutarch *Miscellanies* 2 = A10)

Some seed-like mass breaks off from the boundless stuff and forms into a body in which there is earth at the center, surrounded by layers of air and fire "like bark around a tree." Commentators note the biological analogies such as secretion and growth.[70] The outer layers break away and form into the rings of fire enclosed in air. It is possible that the creative process leaves a residual motion in the heavenly bodies—which is otherwise unexplained.

Anaximander's imaginative model accounts for the apparently circular orbits of the heavenly bodies, at the cost of making them radically different from their manifestations: what we see are disk-like bodies and points of light, whereas what is really up there consists of invisible rings with apertures through which fire is emitted.

---

69. See Bodnár 1992b, who defends the equilibrium account for the earth, and for the heavenly bodies, and Barnes 1982:23–26.
70. See Baldry 1928.

The cosmos exhibits rotational symmetry, which apparently accounts for its stability.

Anaximander extends his naturalistic account of the heavens to the atmosphere immediately around the earth. Thunder and lightning are natural phenomena resulting from wind breaking out of clouds and causing a flash of light and a tearing noise.

Anaximander provides a great deal of precision in his account of the world: we know the shape of the earth, the relative dimensions of its height and diameter, the relative sizes of the heavenly bodies, the geography of the earth, and even the history of the world's development. Yet Anaximander does not give us much in the way of argument for the determinations of his cosmography.[71] What evidence is there that the sun is higher than the moon and the moon than the stars? He seems to assume that there is more fire as one goes farther out in the cosmos, and hence the brightest body is most distant; but he does not argue for this from what we can see. And how is it that we do not see the stars, or their rings, silhouetted against the sun, and particularly the moon? Presumably the brighter body shines right through them.[72] Yet the moon is not obviously brighter than every star. And close observation will show that the moon occults the stars and not vice versa. In geography, how do we know that the continents of the earth are surrounded by an ocean? This problem Herodotus (2.23) would raise later on.

There are also problems for Anaximander's cosmic model. The theory of equilibrium, whereby the earth holds its place because of being equidistant from the heavenly bodies, works well enough in a

---

71. In his biology, he seems to give one argument: land animals including humans were first born from sea creatures, emerging when they were mature; otherwise, they could not nourish themselves (ps.Plutarch *Miscellanies* 2 = A10; Aetius 5.19.4, Censorinus 4.7; Plutarch *Symposium* 730e = A30.
72. Bodnár 1988a argues that only the order Anaximander uses will avoid having, e.g., bands of stars blocked out by the rings of the sun and moon; this is true and important, but it does not solve all problems.

two-dimensional diagram, where we have symmetry. But if we try to envisage a model in three dimensions, allowing for the sun and moon to move obliquely to the earth, then we unbalance the cosmos, and also raise questions about what sort of kinematics would move bodies obliquely to the plane of the earth, if that is a central reference point for the cosmos.[73]

2.2.3 Anaximenes

Anaximander had a student named Anaximenes. We get a convenient summary of his views from Hippolytus:

> Anaximenes, he too being from Miletus, the son of Eurystratus, said the source was boundless air, from which the things that are and were and will be and gods and divinities come to be, the rest from the offspring of this. (2) This is the character of air: when it is very uniform it is imperceptible to sight, but it is discerned by being cold or hot or damp or in motion. It is always in motion; for it would not undergo all the changes it does without moving. (3) For being condensed or thinned it changes its appearance: when it is dispersed to become thinner, it becomes fire; when, on the other hand, air is condensed it becomes winds; and from air cloud is produced by felting; when condensed still more water; when it is condensed even more earth; and when it is condensed as much as possible stones. So the main contraries of generation are hot and cold. (4) The earth is flat riding on air, likewise the sun and moon and the other heavenly bodies, which are all fiery,

---

[73]. This problem has not been faced by defenders of the equilibrium view, including Barnes (partly a defender), Bodnár, and Couprie. On the general issue of the tilting of the heavens, see now Couprie 2009b.

float on air because of their flatness. (5) The heavenly bodies came to be from earth because of the moisture arising from it, which being thinned came to be fire, and from fire floating aloft the stars were composed. There are also some earthy natures in the place of the stars which are carried around with them. (6) He denies that the heavenly bodies move under the earth, as others suppose, but he says they turn around the earth like a felt cap around our head. The sun is hidden not by going under the earth, but by being covered by the higher parts of the earth and by being a greater distance away from us. The stars do not heat us because of their great distance. (7) The winds are generated when air, having been concentrated, is carried along. Being collected and compacted still more clouds are generated and thus turn to water. Hail is produced when water from clouds is frozen as it travels downward; and snow, when more moisture-laden particles get congealed. (8) Lightning is produced when clouds are rent by the force of winds, for when they are rent the flash is bright and fiery. A rainbow is produced when the rays of the sun fall on thickened air. Quaking of the earth is produced when earth is more altered by heat and cold. (*Refutation of All Heresies* 1.7.1–8)

According to Anaximenes, the original stuff of the universe is air, which by being rarefied or condensed turns into other things. When it becomes more rare it turns into fire; when it becomes more dense it turns into wind, cloud, water, earth, and stones, successively. The process of condensation is analogous to that of felting, whereby wool is compacted into felt cloth, and this basic technology serves as a kind of model for Anaximenes.

The earth is flat and stays in place because it is supported by a cushion of air. The heavenly bodies arise as exhalations or evaporations from the earth are further rarefied into fiery bodies; hence they have their own source of light.[74] The sun and moon are like flat leaves, carried around by winds.[75] The turnings, or solstices, of the sun result from the pressure of contrary winds.[76] In speaking of the heavens, Anaximenes uses the analogy of a felt cap turning around above the earth. The stars are like nails fixed to the "ice-like" surface of the heaven.[77] This account could result from a confusion with Empedocles' astronomy (he posits an ice-like firmament), but we should probably accept the general picture because it makes sense of the felt cap imagery.[78] The confusion may only extend to replacing the felt cap image with the image of a crystalline hemisphere. Unlike the sun and moon, the stars maintain a fixed configuration relative to each other, which could be accounted for if they are physically fixed. The heavenly bodies rotate around, not under the earth, and the sun is concealed at night by elevated regions of the earth in the north (see fig. 2.2). This is not good astronomy, but it avoids the physical problem of heavenly rotations interfering with the cushion of air beneath the earth. Doxographical reports assign no account of eclipses to Anaximenes, but one anonymous report may represent his view: a formation of presumably high-level clouds may obscure the sun temporarily.[79]

74. Aetius 2.20.2 = A15, 2.25.2; cf. Aetius 2.13.10 = A14. Theon of Smyrna (198.14 Hiller = Eudemus fr. 145 Wehrli = A16) attributes to Anaximenes, on the authority of Eudemus, the view that the moon gets its light from the sun. Here "Anaximenes" should be emended to "Anaxagoras": Panchenko 2002:324–26; Bicknell 1969:59; Tannery 1930:157. For a new edition of Anaximenes, see Wöhrle 2012.
75. The sun is expressly said to be flat like a leaf (Aetius 2.2.1) and the moon can be inferred to be similarly shaped (cf. Hippolytus *Ref.* 1.7.4). Aetius 2.14.4 is more problematic; see n. 78.
76. Aetius 2.23.1 = A15.
77. Aetius 2.14.3–4 = A14.
78. See Longrigg 1965 for a confusion with Empedocles; Bicknell 1969:54–56 defends the report as authentic. Schwabl 1966 suggests a corrected reading of Aetius 2.14.3–4, followed by Wöhrle 1993: 72; see also an alternative reading proposed by Heath 1913:42; see Mansfeld and Runia 2009:473–74. According to Mansfeld and Runia, the theory that makes the heavenly bodies fiery leaves is anonymous in the Aetius passage; they translate v. 4 as "But some (philosophers declare that they) are fiery leaves, like pictures." But the phrase "like pictures" seems to go best with the nails in the crystalline surface, describing constellations, as others note. In any case there is other evidence that the "leaf" view comes from Anaximenes (see n. 75).
79. See Bicknell 1969:57–65.

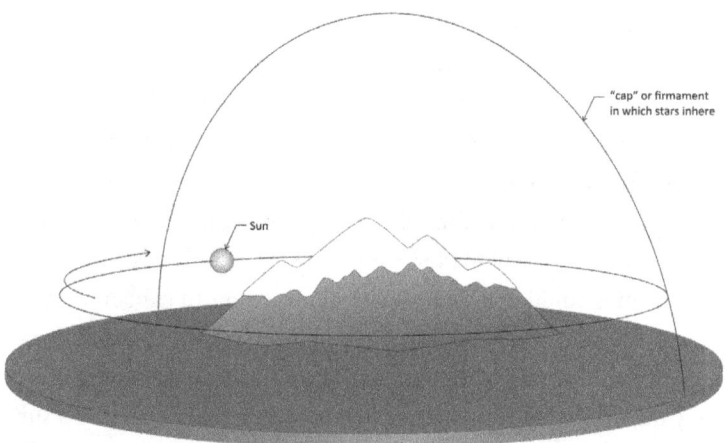

Figure 2.2. Anaximenes' Cosmos. Sun and moon are flat leaf-like structures that circle above a flat earth; the sun "sets" by disappearing behind high mountains to the north. The stars are fixed to a revolving "cap" or dome.

According to one report, "Anaximenes [first discovered] that the moon gets its light from the sun and in what way [the moon] is eclipsed."[80] Although this report goes back to the *Study of Astronomy* of Eudemus of Rhodes, it conflicts with all other reports, according to which the moon has its own light.[81] In any case, if the moon is flat, Anaximenes cannot possibly explain the phases of the moon as caused by shadows.

One problem that remains is a report of earthy bodies, or earthy components of some heavenly bodies. Hippolytus reports "earthy natures" orbiting along with the stars (§5). Most accounts tell us that the sun, the moon, and the stars are all fiery in nature.[82] But the

---

80. Theon of Smyrna *Aspects of Mathematics Useful for the Reading of Plato* 199.1–3, from Dercyllides, reporting on Eudemus of Rhodes = Eudemus fr. 145 Wehrli = 13A16.
81. Bicknell 1969:59; Kirk, Raven, and Schofield 1983:156 and n. 1; Wöhrle 1993:76–77. Tannery 1930:157 suggests that a copyist misread "Anaximenes" for "Anaxagoras." But this interpretation is effectively criticized by Boll 1909, col. 2342.
82. The sun: Aetius 2.20.2; the moon: Aetius 2.25.2, Theodoret 4.23; the stars: Aetius 2.13.10.

earthy bodies are "invisible."[83] Further, ps.Plutarch says "the sun is earth, and because of its rapid motion it gains a very considerable amount of heat" (*Miscellanies* 3, discussed below). This last statement suggests that the sun gets its light and heat from friction rather than combustion, as most other testimonies suggest. Peter Bicknell has tried to defend the earthy nature as a core or nucleus that collects the fiery matter.[84] Pointing out rightly that the earthy dark bodies could not serve Anaximenes either to explain eclipses (since the sun and moon are self-luminous)[85] or to explain meteors[86] (which remained unknown), Bicknell looks for another role they might fill. He argues that the fiery matter needs to have a substance in which to inhere so as to create a unified body. The earthy natures, accordingly, are the nuclei of heavenly bodies. Yet Bicknell solves one alleged problem only to create two others. First, the ancient sources treat the invisible earthy bodies as distinct from the stars. Second, how did the earthy material get into the sky, or, even if it did, how does it stay there? The former question could be accounted for in terms of elemental transformation, such as those used later by Aristotle: some stuff gets compacted by local conditions and condenses into earth. But then how does it remain aloft? We need some force stronger than the winds that blow leaf-like structures about. In any case, the image of the leaf suggests a thin, almost two-dimensional structure. If there is earth in the heavenly bodies, it must consist of a fabric-like layer. Such bodies are not like

---

83. Aetius 2.13.10 = A14.
84. Bicknell 1969:57–71, followed by Wöhrle 1993:70–72.
85. For someone who sees the dark bodies as accounting for eclipses, see Heath 1913:44.
86. In general I shall use "meteor" to designate the phenomena associated with an extraterrestrial body passing through earth's atmosphere. Today astronomers distinguish meteoroids, small solid bodies that can produce such phenomena; larger asteroids; and meteorites, the mineral remains of meteoroids or asteroids that have fallen to earth. (Most meteoroids burn up in the atmosphere without reaching the surface of the earth. Asteroids, by contrast, can fall to earth causing catastrophic damage to the impact site and the environment.)

the heavy, earthy or stony bodies we shall meet at a later stage of cosmology; if there is earth in Anaximenes' heavenly bodies, it plays a subordinate role to fire.[87]

Anaximenes' theory of matter elegantly accounts for meteorological changes. Evaporations from the earth can be condensed into winds, then clouds, and further compacted to take the form of rain. Hail results from the freezing of raindrops, and snow from a different process of freezing. Anaximenes attributes major meteorological changes including the seasons to the action of the sun. He generally follows Anaximander's account of lightning and thunder, except that he seems to think the wind breaking out of clouds is ignited into a fiery burst. Rainbows result from the sun's light falling on dense air. In Anaximenes there is no sharp distinction between heavenly and meteorological phenomena. The same processes drive both, starting with exhalations from the earth. Indeed, his understanding of these processes seems to inform his cosmogony. According to Anaximenes, "When air was felted, he says, the earth was formed first, being completely flat. Therefore it makes sense that it should float on air. The sun and the moon and the other heavenly bodies have their source of generation from earth" (ps.Plutarch *Miscellanies* 3). The earth (that is, the cosmic body, not the stuff) forms first in the cosmos. It provides a source of vapors that form the heavenly bodies and fuel them (Hippolytus §5 above).[88] Thus there is a kind of cosmic ecology in which the presence of the earth in the middle of the cosmos generates the

---

87. Gilbert 1907:688 claims that familiarity with meteorites led Anaximenes to this theory.
88. Wöhrle 1993:19–23, following Klowski 1972 and Hölscher 1953:273–74 uses the ps.Plutarch passage to argue that the traditional account of the ordered transformations of air is wrong: ps.Plutarch shows air changing to earth and earth in turn changing to the fiery heavenly bodies. See also Moran 1973. But the present passage is a cosmogony, not a chemical analysis, focusing on the order in which the heavenly bodies appeared, not the details of elemental transformations. Hippolytus fills in the blanks between earth and fire. See Graham 2003a.

other elements needed to nourish the heavenly bodies, whose heat in turn sustains the generation of those same elements.

2.2.4 Xenophanes

An itinerant poet from Ionia, Xenophanes lived most of his life in exile in Sicily and southern Italy. He is best known for his theological views. But he presented an elaborate theory of natural philosophy and offered explanations of heavenly phenomena. Hippolytus gives a useful summary of his views in this case too:

> Xenophanes of Colophon, son of Orthomenes: he lived until the time of Cyrus. He first said all things were incomprehensible, in these words: [quotes B35.3–4]. (2) He says that nothing comes to be, perishes or moves, and that the totality is one and free from change. He also says that God is everlasting, one, everywhere alike, limited, spherical, and able to perceive in all parts. (3) The sun comes to be every day from tiny flares gathered together, the earth is boundless and surrounded neither by air nor by heaven, and there are numberless suns and moons and everything is from earth. (4) He said the sea is salty because many mixtures flow into it.... (5) Xenophanes thinks a mixture of earth with sea occurs and in time earth is dissolved by the moist, claiming to provide as evidence the fact that sea shells are found in the midst of earth and in mountains; in the quarries of Syracuse impressions of fish and seaweed have been found; in Paros the impression of coral [or: bay] in the depth of a rock; and in Malta fossils of all sea creatures. (6) He says these things happened when all things were covered with mud long ago and the impressions in the mud dried out. The human race becomes extinct when earth is

carried down into the sea and becomes mud, and then the process begins again, and this change occurs in all the world-orders. (*Refutation* 1.14.1–6)

First, Xenophanes' theory of matter. Hippolytus says that "everything is from earth" (§3). We get two sorts of statements from him, one championing earth, one earth and water:

> For from earth are all things and into earth do all things die. (B27)
> All things which come to be and grow are earth and water. (B29)
> For we all come to be [or: are born] from earth and water. (B33)

There was an ancient debate over whether Xenophanes was a material monist for whom the only material substance was earth, because Aristotle said no one held this particular view, some, but not all ancient commentators hesitated to ascribe such a view to him.[89] Some modern scholars see him as a dualist, defending earth and water as twin realities.[90] I am inclined to see Xenophanes as holding the "generating substance theory" with earth as the starting point of transformations.[91] On this view, earth can change to water, water to cloud, cloud to air, and so on, where no common identity is preserved in the transformations, in a law-like and reversible process such as that envisaged by Anaximenes. In any case he seems to recognize many of the transformations, or states of matter on the monist view,

---

89. Theodoret *Therapy for Greek Diseases* 4.5; Olympiodorus *On the Sacred Art* 24; Stobaeus 1.10.12; Hippocrates *On the Nature of Man* 1; disputed by Galen *On Hippocrates' On the Nature of Man* 15.25; cf. Aristotle *Metaphysics* 989a5–6.
90. Lesher 1992:132–33; Finkelberg 1997. Kirk sees in the dualism "a rudimentary physical theory" (Kirk, Raven, and Schofield 1983:176).
91. Graham 2006:70–73.

recognized by Anaximenes: earth, water, cloud, wind, air, fire, and probably, given his views on fossils, stone as well. He clearly recognizes something like a water cycle:

> Sea is the source of water, the source of wind;
> For neither <would there be wind> without great sea,
> nor currents of rivers nor rain water from the sky,
> but great sea is the begetter of clouds, winds
> and rivers. (B30)

Vapors from the sea produce winds and clouds, and by precipitation renew the water on the earth.[92]

Xenophanes' strange view that the earth is boundless and not surrounded by air or heaven (§3) is confirmed by a quotation:

> This upper boundary of earth is visible here at our feet
> touching air; below[93] it reaches down without limit [*es apeiron*].
>   (B28)

Some commentators have envisaged the earth as an infinite column;[94] others see this remark as a limited observation about what can be perceived;[95] one takes the *apeiron* to be other than earth;[96] but it seems from the statement that earth is not surrounded by air or heaven that the surface of the earth is an infinite plane, dividing earth below from air above.[97] Only thus can we understand his view that "there are many suns and moons according to the regions, sections, and zones of the

---

92. Aetius 3.4.4.
93. For the translation, see Mourelatos 2002:332 n. 5.
94. Barnes 1982:27; Couprie 2011:166.
95. Lesher 1992:129–31.
96. Popper 1998:37–39 takes the earth to be bounded by the *apeiron* below it.
97. See Mourelatos 2002.

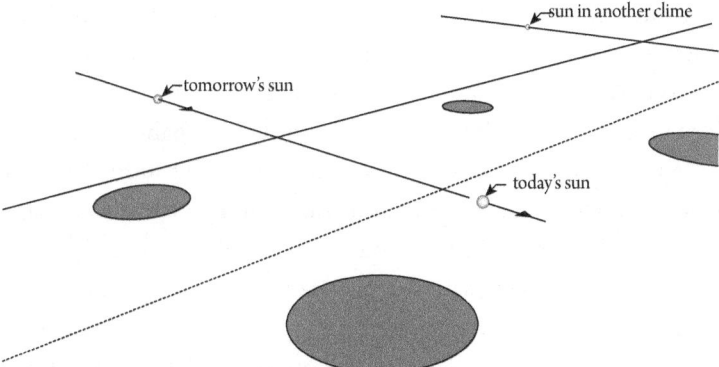

Figure 2.3. Xenophanes' Cosmos. The surface of the earth is an infinite plane separating earth below from air above. Multiple series of suns and moons cross the earth in straight lines (which appear curved from our perspective) in different climes or zones.

earth."[98] Thus Xenophanes' world is more like Thales' than like Anaximander's and Anaximenes' worlds. The dividing line between heaven and earth is an infinite horizontal plane, and heavenly bodies cannot orbit under the earth. And so he said "the sun goes on without end [*eis apeiron*] but seems to circle around because of its distance."[99] If the sun does not circle around from the west, how does it arise in the east every morning? "Another sun in turn comes to be in the east"[100] (see fig. 2.3). Yet he also explains that the sun forms as flares or sparks (*puridia*) gather together, as can be seen on Mt. Ida.[101] Apparently there is some nucleus that continues "on without end," but which can be extinguished

---

98. Aetius 2.24.9; cf. Hippolytus §3. Kirk in Kirk, Raven, and Schofield 1983:174–75 thinks the notion of different suns in different regions is a confusion based on a series of suns visible in one place; and still "Xenophanes permitted himself a certain degree of fantasy" in his astronomy; cf. Guthrie 1962–1981, 1:394 n. 1.
99. Aetius 2.24.9.
100. Aetius 2.24.4.
101. Aetius 2.20.3; on Mt. Ida, Lucretius 5.660–65, Diodorus Siculus 17.7.5–7, Pomponius Mela 1.94–95, with Keyser 1992.

or ignited according to the atmospheric conditions; this nucleus attracts fire particles and unifies them into a diurnal conflagration.

There are multiple moons as there are suns, and each moon is "an incandescent felted cloud" shining with its own light.[102] Thus the moon is presumably more dense than a cloud, but formed from a cloud by a process of condensation. The monthly waning and disappearance of the moon comes about as a result of quenching.[103] As for stars, "Xenophanes [says they come to be] from incandescent clouds; being quenched every day they flare up again at night, like coals; their risings and settings are kindlings and quenchings."[104] These entities seem to have some continuing nucleus, like a coal, which when one blows on it, re-ignites; so the right atmospheric conditions (which presumably are such as to occur in the evening) cause the stars to re-ignite. Perhaps they, like the moon, are felted clouds that can be luminous or dark according to the environment. The stars presumably circle around rather than going on without end, since one can watch many of them slowly moving around the north celestial pole.[105]

Comets and shooting stars are also glowing clouds.[106] Solar eclipses occur when the sun reaches an arid region where it "treads on nothing"; that is, has not enough vaporous fuel to sustain its fire.[107] Similarly, among meteorological phenomena, rainbows result from light falling on dense clouds, and St. Elmo's fire is produced by glowing clouds.[108]

It appears that for Xenophanes, all heavenly phenomena are some kind of concentration of luminous cloud that is fed by exhalations

---

102. Aetius 2.25.4, 2.28.1; on the MS reading of the former and its meaning, see Runia 1989; Mourelatos 2008b.
103. Aetius 2.29.5.
104. Aetius 2.13.4.
105. On astrophysics, see Mourelatos 2008b.
106. Aetius 3.2.11.
107. Aetius 2.24.9 with Bicknell 1967a.
108. See Xenophanes B32; Aetius 2.18.1 with Mourelatos 2008b.

from the earth.[109] Certainly he has the picture of a water cycle in which evaporation rises from the seas to produce clouds, from which rain falls to replace the water lost. The heavenly bodies are a further, more permanent and stable differentiation of clouds, probably existing higher in the atmosphere than the ephemeral cloud formations.

2.2.5 Heraclitus

Diogenes Laertius gives a helpful summary of Heraclitus' physical views:

> Here are his doctrines on particular subjects: Fire is the element and all things are an exchange for fire [cf. B90], as they come to be by rarefaction and condensation—but he explains nothing clearly. All things come to be by contrariety, and the totality is in flux in the manner of a river [B12], and the totality is limited and there is one world-order. It is generated from fire and in turn is consumed by fire in certain cycles, in alternating times through all eternity; and this happens by fate. Of contraries, that which leads to generation is called war and strife [B80], that which leads to conflagration, harmony and peace; and change is a road up and down [B60], and the world comes to be in accordance with it.
>
> (9) As fire is condensed it becomes moist, and coming together it becomes water, and being compacted the water turns into earth. And this is the downward road. Earth is liquified in turn, and from this comes water, and from this everything else, as he attributes almost all phenomena to evaporation from the sea [cf. B31]. And this is the upward road. There are evaporations from both earth and sea, the one bright and pure, the other

---

109. Cf. Guthrie 1962–1981, 1:390–92.

opaque. Fire is increased by the bright vapors, the moist by the others.

What the atmosphere is he does not explain; there are, however, in it bowls with the hollow side turned toward us, in which the bright vapors are collected to produce flames, which are the heavenly bodies. (10) The flame of the sun is the brightest and hottest. The other heavenly bodies are more distant from the earth and for this reason less bright and hot, but the moon being closer to the earth does not travel through the pure region. The sun however lies in a transparent and pure region and maintains a moderate distance from us. Therefore it is hotter and brighter. The sun and moon are eclipsed when the bowls turn up; and the phases of the moon come about as its bowl gradually revolves in itself.

Day and night come about, as well as months, seasons and years, rains, winds, and similar events, as a result of different vapors. (11) The bright vapor burning in the circle of the sun makes day, the contrary vapor as it dominates produces night. And heat increasing from the bright vapor makes summer, moisture growing from the dark brings about winter. In accordance with these principles he gives explanations also of the other phenomena. Concerning the earth he does not declare what its nature is, nor does he explain the bowls. (9.8–11)

Heraclitus is famous as the philosopher of flux (§8), based on the changes of fire, which by rarefaction and condensation (in the manner of Anaximenes) turns into water and earth:

> The turnings of fire: first sea, and of sea half is earth, half fireburst.
>
> <Earth> is liquefied as sea and measured into the same proportion it had before it became earth. (B31 a, b)

The ancient tradition, starting with Aristotle, takes Heraclitus as a material monist whose principle is fire.[110] But in the case of Heraclitus we can establish that he is not a material monist. He says,

> For souls it is death to become water, for water death to become earth, but from earth water is born, and from water soul. (B36)

Here, "soul" is equivalent to fire, and what is crucial is that the death of one "element" is the birth of another: there is no identity between the predecessor element and its successor; there is a causal connection, but no ontological continuity.[111] Whether the elemental changes result from rarefaction and condensation, as Diogenes claims, is unclear from the fragments themselves. He does seem to follow Anaximenes' model broadly, but he does not explicitly speak about mechanisms.[112] Overall, Heraclitus' account of elemental change looks like a simplified version of Anaximenes' account: fire, water, and earth change into each other in an orderly way and according to fixed quantitative ratios: so much fire is equivalent to so much water and so much earth. There is an upward vector from the less rare to the more, and a downward from the less dense to the more.

The alleged conflagration in which the world is consumed periodically by fire and then regenerated out of fire (§8) now seems dubious to most scholars.[113] According to Heraclitus:

> This world-order, the same of all, no god nor man did create, but it ever was and is and will be: everliving fire, kindling in measures and being quenched in measures. (B30)

---

110. *Metaphysics* 984a7–8.
111. See also B76.
112. Simplicius *Physics* 23.33–24.6 = Theophrastus B225 Fortenbaugh et al. But Theophrastus attributed rarefaction and condensation only to Anaximenes: Simplicius *Physics* 149.28–150.4.
113. Kirk 1954:317–24; Marcovich 1967:271–72; Conche 1986:285–86. For a defense of conflagration, see Kahn 1979:134–38, 144–53.

B30 seems to describe an everlasting *kosmos* (the first recorded use of a term meaning roughly "world-order") that is regenerated only piecemeal and gradually ("in measures"), not *in toto* suddenly. Commentators sometimes mistakenly invoke B30 to defend cyclical conflagration, and seem to have little other evidence to cite in its favor.[114]

As for astronomy, Heraclitus seems to have held that the heavenly bodies were like bowls full of fire. The fire is fed by moist exhalations from the earth. The sun burns in clearer air than the moon and hence is brighter. The sun lies between the stars above and the moon below it. Eclipses result from the rotation of the bowls so that the dark exterior is turned towards the earth rather than the bright interior. The same explanation accounts for the phases of the moon, with a slower rotation of the moon's bowl. The model he suggests seems to give the wrong shape, since a bowl is roughly a hemisphere, and as it rotates it should produce a narrowing oval rather than a crescent. But perhaps he thinks of the fire as producing taller flames in the center (where there is the most fuel) and tapering to the rims, thus producing a kind of sphere of flame within the bowl; this would produce the appearance of a spherical body (see fig. 2.4).[115]

Heraclitus does not seem to have explained what the bowls are made of (Diogenes Laertius §11). He may have in mind something like the felted clouds of Xenophanes; in fact we get one late report that says something like that:

> Parmenides and Heraclitus [say] the heavenly bodies are felted from fire [*pilēmata puros*, "condensations of fire" (Mansfeld and Runia)]. (Stobaeus 1.24.1i, Aetius 2.13.8 On the Nature of Heavenly Bodies)

---

114. See Simplicius *On the Heavens* 294.4–7. For a modern defense of the doctrine, see Kahn 1979:134–38, 145–53. This reading of Heraclitus does go back to Aristotle (*On the Heavens* 279b12–17, *Physics* 205a3–4), but his statements are ambiguous. Theophrastus may have misread Aristotle and thus introduced the doctrine of conflagration; see Kirk 1954:317–24.
115. This interpretation was suggested to me by Alexander Mourelatos.

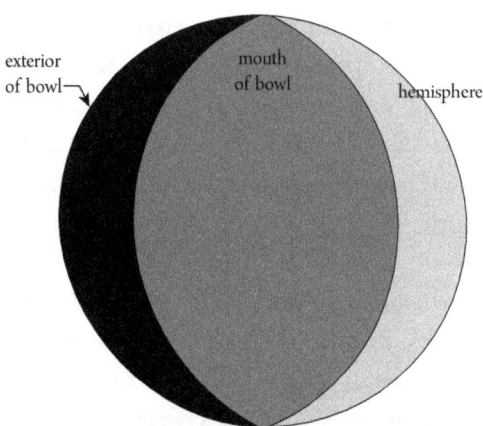

Figure 2.4. Heraclitus' Moon. The moon like other heavenly bodies is a bowl-like structure full of fire. If the fire reaches only to the rim of the bowl, the shape will not approximate, e.g., a crescent or half moon; if, however, the fire creates a hemisphere above the rim, the bowl will approximate the phases of the moon as it rotates.

But what exactly one gets by felting fire (in Anaximenes it is air, in most cases in Heraclitus water) remains unclear here.

Heraclitus says that the sun is new every day.[116] He states that it is the size of a human foot—though it is uncertain whether he is speaking of the actual body or its appearance to us or some other feature of its orbit.[117] The sun will not leave its path, on pain of being punished by the Furies—perhaps Heraclitus' way of saying that a compelling necessity keeps the heavenly bodies in their proper orbits.[118] It is not clear how to reconcile these statements: is there a continuing sun or only a continuing series of similar phenomena?

---

116. B6.
117. B3; Derveni Papyrus IV.6–9, where it is linked with B94. See Lebedev 1985.
118. B94. The fragment may embody a pun: one name for the sun is Hyperion, which could be glossed as ὑπὲρ ἰών (*hyper iōn*), meaning "going overhead" or "going beyond" or overstepping, which will not be allowed (Hussey 1982:55; Gianvittorio 2010:45).

Does necessity constrain the sun or only restrict appearances? Heraclitus did not apparently speak of the shape of the earth or its stability (Diogenes Laertius §11). But presumably he did not advance any innovative theories about it either. There is one account that may preserve a piece of Heraclitus' astronomical theory. An ancient scholiast made this report:

> Heraclitus of Ephesus, a natural philosopher, said that when the sun has come to the western sea and sinks in it, it is quenched, and then traveling through the underworld and reaching the east it is ignited again, and this happens continually. (Scholium on Plato *Republic* 498a)

This suggests that some underground passage (river?) conveys the sun's bowl from one side of the earth to the other. Another report suggests the sun is quenched because of cold in the west and ignited because of heat in the east.[119] Whatever Heraclitus' theory (if he even had a well-developed theory), he seems to have held that the sun's bowl continues in existence, but its contents are consumed and replenished, and its combustion occurs only during the day.

2.3 THE METEOROLOGICAL MODEL

What is perhaps most striking about early theories of the heavens is their diversity and originality. We have heavenly bodies represented as rings, leaves, nails, caps, clouds, and bowls. The early Ionians use analogies to explain the rising and setting of heavenly bodies, the phases of the moon, and solar and lunar eclipses. It seems difficult at first glance to identify anything the several theories have in common

---

119. Olympiodorus *Meteorology* 136.6–11, with Kirk 1954:267–70.

other than the aim to explain the phenomena of the heavens. Yet several features emerge as common to these theories.

(1) All these theories envisage a flat earth. The earth may be floating on a great sea, as in Thales. It may hover in space, as the earth disk in Anaximander. It may ride on a current of air, as in Anaximenes. It may be an infinite plane, as in Xenophanes. But in every case the earth is flat.

(2) The early astronomers share an almost exclusive concern with motions above the earth. In the case of Anaximander the heavenly bodies are wheels vastly larger than the earth. But they must remain equidistant from the earth in order to guarantee his account of the earth's stability. The earth remains in place because of its regular distance from all other bodies. Thus if the heavenly bodies were thought of as skewed relative to the earth, the earth would lose its balance. In his case alone there is some motion below the earth: the stars might possibly (we are not told in detail) move under the earth on their rings. In the case of Thales, we are ill-informed as usual; but if the model of a raft on the sea holds, it would appear that there is no underworld for heavenly bodies to traverse. In the case of Anaximenes, we are told explicitly that the heavenly bodies move horizontally and are hidden by high mountains in the north; they do not travel under the earth, probably for fear that such movement would interfere with the cushion of air which supports the earth.

Xenophanes envisages the earth as an infinite plane, so that all astronomical activity must take place above its surface; below the surface there is only earth without end. Apparently some motions are rectilinear, such as that of the sun and moon, which pass on only to be replaced by other suns and moons. Some motion presumably is circular, such as that of the stars, whose circular paths can be tracked throughout the night. We get no clear picture of Heraclitus' cosmos, but there is no textual evidence that he allowed for a motion passing under the earth.

Indeed, this failure to consider the possibility of heavenly bodies under the earth may be seen as a traditional feature: Hesiod's world consists of a flat, circular earth covered with the dome of heaven above and a chasm of Tartarus below. The heavenly bodies are figures like Helios who drives his chariot across the sky, but does not descend into the dark underworld. Night and Day are personified powers that take turns occupying a house at the edge of the earth by turns, one on the day shift, one on the graveyard shift.[120] The upper and lower realms do not communicate, and indeed guards watch the gates of the lower dungeon to prohibit any unauthorized passage.

(3) The early Ionians seem to account for a limited range of phenomena. They account for the sun, the moon, and the stars. But they do not seem to address the planets (except for a couple of dubious attributions in the doxography),[121] and in any case never identify or name planets.[122] Most of them do not seem to discuss comets. The one who does, Xenophanes, treats them as luminous cloud formations.[123] Planets will become important subjects of ancient astronomy, with one of the main aims of theoretical models being to account for their anomalous motion. But so far, they seem hardly to be noticed. Meteors are not mentioned.[124]

(4) The Ionians treat the heavenly bodies as continuous with phenomena in the lower atmosphere and in some way dependent on them. Heavenly bodies are formed out of exhalations, blown by

---

120. *Theogony* 744–54.
121. Aetius 2.15.6 = DK12A18. Since the opinion is allegedly shared by Metrodorus of Chios and Crates, the latter two may provide the references to planets. Simplicius *On the Heavens* 471.2–6, Eudemus fr. 146 also ascribes knowledge of planets to Anaximander.
122. That is, wandering stars. Technically the sun and moon "wander" relative to the fixed stars, and they are often included in the later list of seven wanderers. Here, however, I am concerned with the five visible planets, Mercury, Venus, Mars, Jupiter, and Saturn.
123. Aetius 3.2.11 = DK 21A44.
124. Aetius 3.2.11 claims that Xenophanes identifies meteors and shooting stars with clouds; but this may just reflect a general attribution of phenomena. In any case, shooting stars need not have been connected to meteors, let alone to meteorites, in this period.

winds or otherwise buffeted by atmospheric conditions, fueled by the exhalations, and liable to be extinguished by a dearth of fuel or a change of conditions. The seasons are typically caused by stiff headwinds that drive the sun away from the northern and southern solstices.[125] In general, they view what we would call astronomical phenomena as indistinguishable from meteorological phenomena.

(5) Furthermore, heavenly bodies are, or can be, ephemeral. For both Xenophanes and Heraclitus the sun is "new every day." Xenophanes' sun can run out of fuel by passing over a desert place, and be extinguished to produce an eclipse. Xenophanes' stars are extinguished by day and ignited by night, as apparently are Heraclitus'. This feature is appropriate for phenomena that are essentially meteorological in character.

(6) Heavenly bodies are in general as light as or lighter than air. Anaximander explains the stationary position of the earth by its equilibrium with heavenly bodies. But he has no particular worry about the sun, moon, and stars: they are composed of rings of fire surrounded by air. These bodies are not heavy and need no particular explanation for the fact that they do not fall. Anaximenes develops an account of matter that seems to influence early physical speculation: it is by rarefaction or condensation that the basic substance, namely air, turns into other substances, such as (by rarefaction) fire or (by increasing condensation) wind, cloud, water, earth, and stones. For a model he uses felting, a process by which wool is compacted to form felt. For him the heavenly bodies are felted, but they do not necessarily become very dense. The sun and moon are blown around like leaves floating on air. For Xenophanes the heavenly bodies are kinds of cloud that naturally float above the earth. Heraclitus has bowls full of fire. We do not know what the bowls are composed of, but since they only contain fire, they need not be made of any very dense

---

125. Aetius 2.23.1.

material. Accordingly, there seems to be no major problem for sixth-century physics about why the heavenly bodies stay aloft: they are light, and in many cases lighter than air.

(7) The heavenly bodies have their own source of light. For Anaximander the heavenly rings are filled with fire. For Anaximenes the heavenly bodies are essentially fiery. For Xenophanes they are fiery or luminous clouds. For Heraclitus they are bowls of fire fed by moist vapors from the earth. They do not need to get their light from some outside source: they exhibit a property of self-illumination.

(8) There is no agreement on the shapes of heavenly bodies. They may be ring-like or wheel-like (Anaximander), thin and leaf-like (Anaximenes), cloud-like (Xenophanes), or bowl-like (Heraclitus).

(9) The source of their motion is various and often unclear. They may move by a residual motion from an original explosion (Anaximander), or by winds (Anaximenes, Xenophanes). But in general there seems to have been little discussion of the motive force connected with the daily and yearly motions of the heavenly bodies.

In general, the astronomy of the Ionians deals with mid-air phenomena above the earth, what the Greeks call *meteōra*. Unlike modern astronomy, and even Aristotelian astronomy, there is no sharp dividing line between atmospheric conditions and astronomical events. There is a continuity between the lower atmosphere and the upper regions, what would sometimes be distinguished as the region of air and the region of aether. For Aristotle there is a sharp division between the sublunary world, with it four elements in everlasting flux and its seasonal cycles (in what is now called the troposphere), and the heavens proper where there is no change other than ceaseless, regular, circular motion in a realm composed of the fifth element, known vulgarly as the aether (*aithēr*). For the early Ionians, by contrast, there is a continuous realm of atmospheric vapors and exhalations where luminous bodies float or are blown, and which depend on vapors for their fuel. Some authors seem to view the heavenly bodies as feeding on vapors like cattle grazing in a pasture, in what one scholar calls the "pasture-theory."[126]

Aristotle seems to be criticizing precisely this view when he says:

> Thus everyone before me who supposed the sun was nourished by the moist was being ridiculous. And for this reason some claim the sun makes its turnings (solstices), for the same places cannot always supply its nourishment. . . . Just as ordinary fire burns as long as it has fuel, and the moist is the only fuel for fire, so, they assume, the rising moisture reaches as far as the sun, and its rising is like that of ordinary flame, as the basis for their explanation. (*Meteorology* 354b33–355a8)

Similarly, a Hippocratic treatise understands heavenly bodies as receiving fuel from the atmosphere:

> The path of the sun, moon, and stars is through the air [*pneuma*]. For air is the fuel [*trophē*] of fire, and fire deprived of air would not be able to survive. So thin air supports the everlasting life of the sun. (Hippocrates *On Breaths* 3)

Other writers reflect the same ecological interpretation of the heavens.[127]

A view of the heavens and astrophysics that would later come to be seen as naive and obsolete constituted the standard view in the sixth century. Vapors from the sea supplied the heavens with air,

---

126. Stokes 1962–1963, 2:6–8. He distinguishes the pasture-theory from the view that has the wind pushing the sun at the solstices, but for my purposes the general characterization is more important than the detailed physics of solstices. See the echo in the epigraph to this chapter.
127. Lucretius 1.1090, 1.231; Herodotus 2.25.1–3.

wind, and cloud; they fueled the fires seen in the sun, moon, and stars. Weather patterns influenced and perhaps drove the heavenly bodies, causing seasonal phenomena such as turnings, and occasional phenomena such as eclipses, shooting stars, rainbows, and St. Elmo's fire. The explanations were thoroughly naturalistic, but they drew no strong distinctions between meteorology and astronomy, between the world of change and the changeless spheres, between the sublunary and the superlunary, the earthly and the cosmic. I shall accordingly designate the view they adhere to as the Meteorological Model.

Chapter 3

Borrowed Light

The Insights of Parmenides

> A light by night, wandering around earth with borrowed light.
> —Parmenides B14

3.1 FIFTH-CENTURY ADVANCES

In the first half of the fifth century BC we suddenly meet two theorists who seem to have made major advances in understanding astronomy. Both Anaxagoras (c. 500–428) and Empedocles (c. 495–435) understand that the moon gets its light from the sun. They also understand the causes of both solar and lunar eclipses: in a solar eclipse the moon intervenes between the sun and the earth, blocking the sun's light to the earth; in a lunar eclipse, the earth intervenes between the sun and the moon, blocking the sun's light to the moon. These theories appear in testimonies of the respective philosophers, and they are supported by several verbatim fragments. We are, then, on firm ground in attributing the theories to these philosophers.

To be sure, the theories of Anaxagoras and Empedocles are not yet modern theories. Both have a world with the earth at the center rather than the sun. Anaxagoras certainly and Empedocles probably have a flat earth. Anaxagoras allows for heavenly bodies that are normally invisible to us also to cause lunar eclipses. Furthermore, Anaxagoras is said to have "predicted" the fall of a meteor at Aegospotami in northern Greece. In fact such a meteor fell around 467/6 BC and became famous throughout the Greek world. Since, as we know, meteors cannot usually be predicted,[1] this report raises questions similar to those raised about Thales' prediction of a solar eclipse.

How, then, did Anaxagoras and Empedocles arrive at their theories? Why did these students of the heavens come upon the right theories in the first half of the fifth century when others did not (apparently) do so in the sixth? Moreover, why did they come to this understanding centuries before other intelligent and often more observant astronomers in other cultures which kept records about the same time, notably the Babylonians, Egyptians, Indians, and Chinese?

The early fifth century seems to provide something like a major scientific advance in astronomy. How scientific it is can be determined only by understanding how it came about: was it the product of a scientific process of induction, or theory construction and testing, or conjectures and refutations? Or did it come about as the product of some unscientific sequence of ideas or unsubstantiated conjectures? If both Anaxagoras and Empedocles discovered the true nature of some astronomical phenomena, who was first, or did they make the discovery independently, or in conjunction with each other? Let me present a series of questions, not all of them independent of each other, but conceptually distinct, which will occupy the remainder of this study:

1. Who discovered the theories in question first?
2. What led him to this discovery?
3. Did Anaxagoras and Empedocles have good evidence for their astronomical theories?

---

1. There are some meteor showers that arise when the earth passes through the trail of a comet in its orbit around the sun. It is now possible, with the aid of telescopes and computers, to predict the near miss or hit of an asteroid. But these techniques presuppose knowledge and instruments unavailable to the ancients.

4. Did the community of philosophers accept the theories?
5. Did Greek theorists of the fifth century develop the correct theory of eclipses on their own, or did they borrow it from another source (Thales, Pythagoras, the Babylonians)?

The first question is a historical and biographical one. The second a theoretical and methodological one. The third an epistemological one. The fourth a social one. And the fifth is a question of originality. All these questions are in some measure empirical: whatever notions we may have of early Greek philosophy and of astronomy, it is a factual matter how philosophers confronted the astronomical phenomena they studied, how they articulated the problems, and how they solved them. Inevitably, the data on which to build an interpretation are scanty because so little documentary evidence has survived; we have no journals, archives, letters, laboratory or research notes, newspaper reports, interviews, or other sources such as that from which modern history of science can be constructed. Nevertheless, there is some evidence in the historical record, and to it we must turn to try to reconstruct the stages by which the new understanding of the heavenly bodies arose.

3.2 THREE INSIGHTS: HELIOPHOTISM, PLANETARY UNIFICATION, SPHERICITY

In the advances of Anaxagoras and Empedocles we can identify two distinct astronomical theories. One is a theory about the moon's light. Previous natural philosophers had all attributed the moon's light to the moon itself: it had the property of being self-luminous, as was seen in ch. 2. It was fiery, or glowing in its own right. According to Anaxagoras and Empedocles, however, the moon depends for its light on external source. That source is the sun. We can characterize the first kind of theory as ascribing *idiophotism*, or self-illumination, to the moon, the second as ascribing *heliophotism*, or solar illumination.[2]

The second theory is about eclipses. Eclipses are caused by the blocking of the sun's light, either by the earth, or by the moon, or perhaps by some other heavenly bodies. Let me call this theory *antiphraxis* after the Greek (Aristotelian) word for "blocking," "screening," or "interposition." Earlier philosophers had provided interesting theories of eclipses, but none had offered antiphraxis as an explanation.

Now these two theories are related in a certain way. Heliophotism does not of itself entail antiphraxis. For it might be the case that, for instance, the moon gets its light from the sun, but the sun loses its light because it runs out of fuel, as Xenophanes thought, or the moon suffers eclipse because a cloud covers it, as Anaximenes seems to have thought. On the other hand, antiphraxis presupposes heliophotism because it accounts for lunar eclipses by having the earth (or another body) block the moon's source of light, and it accounts for solar eclipses by having the moon function as a dark body during the new moon phase (as we shall see) to screen the sun's light. Thus heliophotism is a necessary condition for antiphraxis and antiphraxis a sufficient condition for heliophotism. In a certain sense, antiphraxis exploits the insights of heliophotism to explain eclipse phenomena. It builds on heliophotism as a foundation.

Heliophotism, accordingly, could be the key to whatever advances occurred in the first half of the fifth century. It is a striking doctrine that seems to open up other possibilities for explanation. Where did this theory come from? One way of tracking it is to go back to the doxographical tradition that grew up around the research of Theophrastus. Theophrastus wrote a study, *Doctrines on Nature*, in sixteen books, which summarized and perhaps criticized

---

2. Alexander Mourelatos coined the latter term; I am coining the former (which builds on the terminology of the doxographers).

his predecessors' views on natural philosophy.[3] Too long to be easily transmitted but too valuable to be ignored, this work became the basis of numerous epitomes and digests that we know as doxography, surveys of opinions. Around the first century of the Christian Era Aetius made a further epitome.[4] He records the following doctrine about the sources of the moon's light:

> Thales first said it [the moon] is illuminated by the sun. Pythagoras, Parmenides, Empedocles, Anaxagoras, Metrodorus likewise. (Aetius 2.28.5)

This report, perhaps going back to Theophrastus, cites Thales, Pythagoras, and Parmenides as holding heliophotism before Empedocles and Anaxagoras (since Metrodorus came later).

We have already discussed Thales. It is doubtful he held this view, but if he did, he did not manage to convince anyone else of it. Even if Thales was ultimately responsible for heliophotism, we would need some sort of evidence showing that his very own theory became influential in the early fifth century after being ignored in the sixth—evidence that we do not have.

Pythagoras is a new figure in this story. As we noted above (ch. 1.3) Pythagoras or the Pythagoreans have often been viewed as the originators of the best ideas in early Greek astronomy. Starting with Paul Tannery and John Burnet, historians of science and philosophy have seen Pythagoras and his school as developing a mathematical astronomy that recognized the source of the moon's light and the causes of eclipses.[5]

---

3. Diogenes Laertius 5.48.
4. On Aetius see Mansfeld and Runia 1997.
5. Tannery's *Pour l'histoire de la science hellène* (1887, 2nd edn. 1930) and John Burnet's *Early Greek Philosophy* (1892, 4th edn. 1930) assigned a major role to Pythagoras and the Pythagoreans in science and philosophy, respectively. This view was influential, especially in the first half of the twentieth century. See, e.g., Cornford 1939, Raven 1948.

But recent scholarship has established that we have no reliable scientific cosmology from him. This is a complicated question, but fortunately much helpful work has been done, and the evidence for a contribution by Pythagoras is almost nil.[6] What can be confidently attributed to Pythagoras is the doctrine of reincarnation. The only systematic cosmology we have for a Pythagorean is that for Philolaus, who wrote toward the end of the fifth century, that is, after Parmenides and many of the other figures associated with Greek cosmology.[7] To infer an early Pythagorean cosmology from what is found in Philolaus and from alleged reactions to an unknown theory in other cosmologists is to conjure up a theory from surmises and wishful thinking. Unfortunately, Philolaus' theory seems to become the default "Pythagorean" theory (Aristotle *On the Heavens* 293a20–24) that ancient sources project back to Pythagoras. It is best to take Philolaus as reacting to cosmologies in the Ionian tradition.[8] Books continue to be written about Pythagoras and his tradition that make optimistic claims.[9] But to date the claims of an early Pythagorean philosophical and cosmological system remain unsubstantiated.[10]

This brings us to Parmenides. We know Parmenides as a critic of natural philosophy, who on most accounts dealt Ionian philosophy a blow from which it never fully recovered. Thus it seems implausible to take him seriously as a source of the doctrine. And yet, we have evidence for this very doctrine in his own words. Parmenides finished his philosophical poem with a cosmology of his own, which he

---

6. Burkert 1972:299–368; Huffman 1993. See also earlier criticisms in Frank 1923; Cherniss 1935:384–98; Heidel 1940; Vlastos 1953.
7. Huffman 1993:239.
8. Cf. Huffman 1993:240–41.
9. Zhmud 1997; Kahn 2001; Riedweg 2005.
10. The difference between previous and current views on the Pythagoreans are reflected in many of the changes between Kirk and Raven 1957 and its successor Kirk, Raven, and Schofield 1983. Huffman 2013 argues that some of the *akousmata* of the Pythagorean tradition could reflect Pythagoras' views on cosmology; but they show a mythological rather than a scientific understanding of heavenly manifestations.

called "deceptive," but which he claimed was superior to that of any of his predecessors.[11] The cosmology included astronomical theories, and among them a treatment of the sun and moon:

B14. [of the moon] a light by night,[12] wandering around earth with borrowed light,

B15. ever gazing toward the rays of the sun.

I present these two fragments as continuous, though they are not; but their message goes together well, with B15 being, like B14, a treatment of the moon in relation to the sun. Here B15 offers empirical evidence for the assertion in B14. Since the moon's shining face is always turned towards the sun, we can see that it borrows its light from the sun. In other words, the luminous part of the moon is reflecting the sun's light that falls on it.[13]

In poetic terms and apparently without fanfare, Parmenides offers a stunning insight: the moon is a body lacking its own source of light. It gets its light from the sun, as we can see by tracking its phases, noting that the luminous part is always facing towards the sun. This insight describes the actual state of affairs between these two heavenly bodies. To our knowledge, it had never been understood previously by anyone in any other culture.[14] Parmenides, the alleged anti-cosmologist gets it just right for the first time ever. To be sure,

---

11. B8.51–52, 60–61.
12. Reading νυκτὶ φάος with MSS. Most editors follow Scaliger's emendation to νυκτιφαὲς. But in favor of the MS reading, see Mourelatos 2012, who also argues that this fragment proposes heliophotism.
13. Guthrie 1962–1981, 2:66 doubts that these lines show an understanding of lunar light. (See earlier Tannery 1930:216–19 for a stronger rejection, followed by Heath 1913:75–77.) Yet it is clear from his imitation of Parmenides' lines that Empedocles understands them as having the right implications, as will be pointed out in the next chapter: see Empedocles B45 with B43, B47; and Coxon 1986:245. See discussion of Karl Popper's view below.
14. See Rochberg 2004:278. Beatty 1997/1998 claims that the Egyptians made the discovery in the New Kingdom (1600–1080 BC), but the evidence seems to me to be dubious at best.

scholars have sometimes claimed that Parmenides must have gotten this idea from someone else, often an unnamed Pythagorean.[15] But why? In every other case when a philosopher is the first to propound a theory, we tend to accept it as his own. In Parmenides' case we posit an unknown Pythagorean. Obviously it is difficult to accept the fact that Parmenides is anything but an ivory tower theoretician. What we can say is that of all the philosophers theorizing before him, none had this insight. It is possible he borrowed this insight from some anonymous scientist; but by Ockham's razor we seem justified in eliminating this hypothesis until such time as someone provides positive evidence for it.[16]

Furthermore, Parmenides, we are told, is responsible for two other astronomical insights. He first identified the Morning Star with the Evening Star.[17] And he first held that the earth was spherical.[18] Now, as has been mentioned, the identification of the morning star and the evening star had been accomplished in Babylonia more than a millennium before Parmenides. Thus Parmenides cannot claim to be the first discoverer without qualification. But was he the first Greek to discover this? The only planet(s) mentioned by Homer, Hesiod, and other early writers are the morning star and evening star.[19] In fact we find little of planets in early philosophical writings.[20] In Eudemus and Aetius, we find

---

15. Tannery 1930:219. In a twist on this story, Popper 1998:100 suggests that Parmenides made his astronomical discoveries when he was himself still a Pythagorean. On Parmenides as a Pythagorean, see Diogenes Laertius 9.21 = 28A1.
16. See Wöhrle 1995.
17. Aëtius 2.15.7 from Stobaeus 1.24.2e = 28A40a; Diogenes Laertius 9.23. The first passage is quoted below, section 3.5.
18. Diogenes Laertius 8.48, 9.21 = 28A1; this attribution is rejected by Morrison 1955:64.
19. Homer *Iliad* 22.318, 23.226; *Odyssey* 13.93–4; Hesiod *Theogony* 381; Pindar *Isthmian* 4.24. A scholiast names the poet Ibycus (sixth century BC) as the first to combine the names Dawn-bringer (*heōsphoros*) and Evening Star (*hesperos*) (Ibycus fr. 331, scholium on Basil *Peri geneseōs*), but provides no further evidence or context for the claim.
20. Dicks 1970:46, 47, 58; Bicknell 1968a; West 1980. "As for the planets, it is doubtful whether they were recognized as such before the fifth century B.C. . . ." (Dicks 1970:30). See also Burkert 1972:310–11.

opinions about planets attributed to only one early writer, Anaximander, and this attribution is dubious.[21] Since Aetius does report (from Theophrastus) surveys of evidence on heavenly bodies, we seem to have not just an absence of theory, but evidence that there was no theoretical reflection on planets. Anaximenes may distinguish between the fixed stars, which are attached to the cap-like firmament of heaven, and the planets, which move freely.[22] But here the "planets" may designate only the sun and moon, which are technically wandering bodies whose movement does not conform to that of the fixed stars. The first promising mention is that of Alcmaeon, who is said to have noticed that planets (sometimes) move contrary to the east-west course of the fixed stars.[23]

We get one intriguing mention of a named planet, *Mesonux*, the Midnight or Nighttime Planet, which seems to go back to the poet Stesichorus (a contemporary of Thales).[24] The name implies that the planet was visible late at night, evidently in contrast to the planet(s) visible only at dawn or dusk, namely the Morning and Evening Stars (i.e., Venus, as well as Mercury, much less visible). Since, however, there are three nighttime planets: Mars, Jupiter, and Saturn, the level of observation cannot have been great in the sixth century; no one seems to have taken the trouble to differentiate and track different planets, other than to distinguish between crepuscular and midnight apparitions. Yet at least we see that the Greeks of this era were starting to pay attention to the planets. As one scholar says, "There are many signs of an increased

---

21. Simplicius *On the Heavens* 471.2–6, Eudemus fr. 146; Aetius 2.15.6. With Anaximander Aetius joins Metrodorus of Chios and Crates, and hence may conflate their later accounts of planets with an earlier account that does not refer to them. In the chapter that we would expect to find information about planets, 2.13, we find nothing for the early Presocratics; see reconstruction in Mansfeld and Runia 2009:466–67.
22. Heath 1913:42–43. To the contrary, Dicks 1966:30; Dicks 1970:46–47. Burkert 1972:311 gives a balanced view. There is a question of how to understand Aetius 2.14.3–4; see Mansfeld and Runia 2009:474–75. See above, ch. 2.2.3.
23. Aetius 2.16.2–3 = 24A4.
24. Herodian 1.45.14; 2.743.24; Bicknell 1968a; West 1980.

theoretical interest in planets in the second half of the sixth century and the first half of the fifth...."[25] In general, the state of knowledge in the early fifth century, as far as can be seen from later reports, seems consistent with the statement that Parmenides was the first to identify the morning star and the evening star—in the Greek tradition.

The first mention of the five visible planets among Greek philosophers and astronomers seems to occur in Philolaus, in the last third of the fifth century.[26] The fact that he has five and not, for instance, seven (two morning and two evening stars—and he distinguishes the sun and moon from the planets), shows that by then the identity of the inner planets Venus and Mercury had been recognized. But this point has immediate ramifications for the relationship between Greek and Babylonian astronomy. If the Greeks had known *anything at all* about Babylonian astronomy, for instance in the time of Thales, they would have known that there are five and only five visible planets. This fact had been built in to all Mesopotamian astronomical tables and almanacs for centuries before Thales.[27] The fact that the most educated and advanced Greeks had to rediscover the number of the planets suggests that they did not have any significant contact with Babylonian planetary theory and observation. In the later fourth century, the pseudo-Platonic *Epinomis* shows acquaintance with the Babylonian planets, and Aristotle praises the Babylonians and Egyptians for their careful observations.[28] But in the period of the Presocratics what

---

25. West 1980:207. He adds, "they are treated, so far as we can see, as an open class rather than as distinct individuals."
26. Aetius 2.7.7 from Stobaeus 1.22.1d = A16. Philolaus also has one invisible planet, the counter earth. See Dicks 1970:72; see Huffman 1993:240–61 for a plausible and sympathetic reconstruction of his system.
27. The MUL. APIN reached its final form perhaps around 700 BC, but drew on texts that were a few centuries earlier; it tracks the five known planets. See Hunger and Pingree 1999:57, 73–75. The *Enūma Anu Enlil* goes back almost a millennium to observations from the reign of Ammisaduqa in the seventeenth century BC; its tablet 63 studies omens connected with the appearances of Venus, whether morning star or evening star. See Reiner and Pingree 1975.
28. Ps.Plato *Epinomis* 987a–d; Aristotle *On the Heavens* 292a7–9; see Lloyd 1979:176–80.

is striking is the absence of any confirmed borrowing from a neighboring tradition that was far in advance of the Greeks in identification and observation of the wandering stars.

It seems significant, too, that Parmenides is far removed from contact with the Persian Empire in Italian Elea. And his discovery comes precisely at a time when contact between the Greeks and the Asians was most difficult. The Persians sent an expedition to punish the Greeks for raiding their territory, resulting in the Battle of Marathon in 490. After this loss, the Great King of Persia spent years organizing a massive invasion of Greece, culminating in the battles of Salamis and Platea in 480 and 479. After winning unexpected victories, the Greek defenders went on the offensive with raids and conquests on the coastline of Asia Minor. In other words, throughout this period regular and peaceful cultural exchanges between the Persian Empire and the Greek city-states were difficult or nonexistent. It seems gratuitous to suggest that after the Greeks had gone on in ignorance of Babylonian advances for decades when peaceful communications were possible, they suddenly came into contact with their ideas when communications were difficult or impossible.

More striking still is Parmenides' alleged insight that the earth is spherical:

> He was the first to say the earth was spherical and situated in the middle. (Diogenes Laertius 9.21)

The same view appears rather suddenly in Plato's *Phaedo* (ca. 485–480 BC), where its provenance is not discussed.[29] Aristotle defends and indeed proves the sphericity of the earth in the *On the Heavens* a few years later with adequate scientific arguments.[30] Philolaus' astronomical

---

29. *Phaedo* 109a; Dicks 1970:95–96.
30. *On the Heavens* 2, 297a8–298a20; Dicks 1970:197–98.

theory that makes the earth a planet orbiting around the central fire of the world envisages the earth as spherical. But Parmenides (apparently) has a central earth, and all his predecessors have some version of a flat earth. How does he come to such a view, or is the report unreliable? A recent study by Dmitri Panchenko has shown that Anaxagoras argued against a theory that the earth is spherical.[31] Who could be the object of his polemic but Parmenides? Thus we have historical corroboration for this remarkable insight.[32]

Accordingly, the three astronomical insights attributed to Parmenides seem to be supported by independent evidence: Parmenides' own words in the case of heliophotism, a consistent historical development in the case of the planet Venus, and historical counter-argument in the case of the sphericity of the earth. We seem justified in taking Parmenides as a serious astronomer. Indeed, his three insights suggest that he is by far the most insightful and powerful astronomical theorist of all the early philosophers; for as ingenious as his predecessors were, none made any lasting contribution to astronomical theory as did Parmenides. Within a century and a half, all of his insights would be adopted by astronomers and vindicated by empirical observations made in the framework of improved explanatory models. Whatever position we take about the role of Parmenides' cosmology in relation to his philosophy, we need to stop ignoring or patronizing him as an astronomer and scientist. The most remarkable burst of creative energy in early Greek astronomy begins with the speculations of Parmenides. The greatest ontologist and most abstract philosopher among the Presocratics is also the best theoretical scientist of his time.[33]

---

31. Panchenko 1997 with Aristotle *On the Heavens* 293b33–294a4, Martianus Capella 6.590, 592.
32. See also Kahn 1960:115–18, followed by Guthrie 1962–1981, 2:64–65.
33. While most scholars still view Parmenides primarily as a philosopher and perhaps in particular as a metaphysician, there are those who now see him as an important contributor to science. See Cerri 1999; Bollack 2006; Cordero et al. 2008; and others who see his work as compatible with scientific theories, e.g. Curd 1998, Palmer 2009. I will speak later about Popper 1998.

3.3 THE POWER OF A MODEL

How did Parmenides arrive at the theory of heliophotism? We have no documentary evidence at this point and can only conjecture. One interesting suggestion has been made by Alexander Mourelatos.[34] Xenophanes, Parmenides' closest predecessor, viewed heavenly bodies as cloud formations—perhaps more lofty and more stable, but still continuous with the mid-air phenomena of clouds. We can, of course, view clouds from much closer than heavenly bodies, and we can observe that they seem continuous with exhalations and mists from standing water, steam from boiling kettles, and fog. These phenomena have no light of their own, but are commonly illuminated by the sun. We can observe dark shadows under thick cumulus or cumulonimbus clouds, highlights on their tops, and so on. In the evening, the setting sun makes them glow yellow or red. The luminous portion of the cloud faces the sun, the dark side faces away. At times, as Mourelatos points out, we see the moon during the daytime in the vicinity of clouds. Sometimes the color and texture of the moon seem very similar to those of clouds. For someone taught to look for similarities between heavenly bodies and clouds, the similarity is striking and could be highly suggestive: just as the cloud derives its light from the sun, so does the moon. The similarity in appearance between clouds and moon indicates a deeper similarity in their common dependence on an external source of illumination.

While we cannot confirm this starting point, the story does fit the historical situation and show how one philosophical theory can lead to another. Whatever the precise inspiration for Parmenides' insight, he came to see that the moon might require illumination from another body. The beauty of heliophotism is that to hypothesize it is in a sense to prove it. The first point is that made by Parmenides B15: the

---

34. Mourelatos 2002b; cf. Mourelatos 2008b:151–52, and earlier Bicknell 1967b.

shiny part of the moon is always facing the sun. The most striking fact about the moon is that its phases are constantly changing: the "inconstant moon" presents a different aspect from day to day, finally disappearing at the end of the lunar month for two to three days, and then reappearing. But one thing is constant: the luminous part of the moon is always facing the sun.

Armed with this preliminary observation, let us follow the moon through one month (see fig. 3.1). At sunset the sun disappears on the western horizon. On the first day of the lunar month, a thin rim of light stands suspended low in the western sky, above the sun and slightly to the left (east) at a diagonal (at a latitude similar to that of

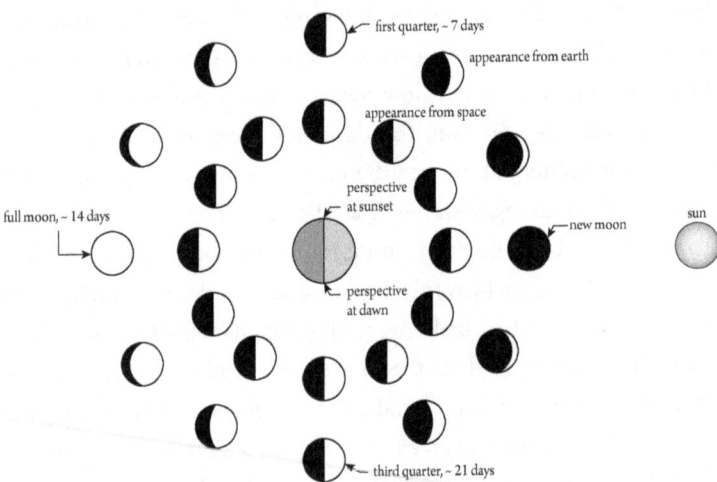

Figure 3.1. The Positions and Phases of the Moon through a Lunar Month. The moon revolves eastward around the earth in a lunar (synodic) month of approximately 29.5 days. Because of the earth's rotation, the moon appears to circle the earth daily from east to west, slightly slower than the sun, thus gradually moving eastward relative to the setting sun in the evening and presenting more of its illuminated surface (waxing crescent and gibbous phases) until the full moon; after which it moves eastward toward the rising sun in the morning, presenting less of its illuminated surface (waning gibbous and crescent phases).

Greece or southern Italy, where Parmenides lived). As the heavens rotate westward the moon sinks below the horizon in pursuit of the sun. The next evening as the sun sets the moon is still a thin crescent, slightly fuller than the previous night, and slightly higher in the sky and farther east. The following evening the crescent is fuller again, slightly higher than the previous night and farther east. And so it goes until after about a week the moon is roughly in the south, presenting about a half-circle of light, with the luminous half facing the setting sun. The following night the moon is in the southeast quadrant and more than half full, a gibbous moon with the luminous portion still facing the setting sun. Finally in the middle of the month after about two weeks, the moon is just rising as the sun sets, and it is a full moon, 180° removed from the sun.

The following night the moon does not rise until the sun has set, and its luminous part is facing east rather than west. But now, we realize, it is closer to the rising sun of the next morning than the setting sun of the evening. So we track it with relation to the rising sun. The same phases we observed in the first half of the lunar month now repeat backwards, with the moon getting closer to the rising sun each night, becoming thinner in appearance until it shrinks to a half moon about a week later, standing in the south as the sun rises in the east. The moon now becomes a waning crescent night by night until it disappears into the vicinity of the sun.

All of this is routine astronomy that can be followed in any almanac. No doubt it was familiar to any Greek who watched the moon, as probably most Greeks did, since their calendar was based on lunar months. But what is striking is that on the basis of the insight that the moon gets its light from the sun, every sighting becomes empirical evidence for the hypothesis. The moon's shape is a *function* of its angular distance from the sun, and its luminous part is always facing the sun. This is what heliophotism, taken as a hypothesis, predicts. The phases of the moon present not just repeating patterns, but *empirical evidence* for a scientific hypothesis. If the hypothesis is right, the appearance of the moon is not just an accidental correlation of its position relative to the sun, but a *result* of that position. If the sun is the source of the moon's light, the moon could not look any other way than it does. In one month's time we have turned the heaven into a laboratory and confirmed a causal relation between the sun and the moon. A relationship that every human being had observed since time immemorial, and no one of whom had understood as a causal relation. Not bad for a speculative philosopher.

What is crucial here is that once one has tracked the lunar phases and evaluated them on the basis of a bold explanatory hypothesis, one will probably never be able to see the relationship between the moon and the sun in any other way. The appearances suddenly make sense in a way they never did before. The moon wanders around the earth "with borrowed light" (B14, quoted above). Parmenides uses the word *allotrion phōs*, creating a word-play with the Homeric phrase *allotrios phōs* "foreign man."[35] He creates a poetic tour-de-force to express his scientific tour-de-force. The moon is a dependent.

3.4 CONJECTURES

Parmenides' contributions to cosmological theories have mostly been ignored or downplayed by commentators. But there is one major exception. The eminent philosopher Karl Popper became interested in Parmenides' astronomical theories toward the end of his life and wrote several essays on the subject.[36] According to Popper, Parmenides made a significant scientific advance in his understanding of astronomy, which in fact informed his whole cosmological

---

35. Homer *Odyssey* 18.219.
36. Collected in Popper 1998, essays 3–5.

theory. Thus in some ways Parmenides the philosopher was indebted to Parmenides the astronomer.

Popper begins by recognizing Parmenides as a significant cosmologist, on the basis of his empirical discoveries. This gives rise to a problem he calls the "recoil from sensualism," namely, "How is it possible that a successful astronomer and empiricist can turn radically against observation and the senses, as Parmenides did in his Way of Truth?" (Popper 1998:80, italics removed). There is a second problem, of how Parmenides could say the world of opinion did not exist, which Popper thinks is both paradoxical and anachronistic (insofar as it anticipates Kant). The solution to the first problem derives from his discovery.

> Parmenides discovered that the observation ... that the Moon—Selene—waxes and wanes during the course of time is false. Selene does nothing of the kind. She does not change in any way. Her apparent changes are an illusion. ... [T]he discovery that the Moon neither waxes nor wanes was in its turn made with the help of observation. ... So observation may imply the falsity of observation. ... The apparent bodily change[s] of the Moon turn out to be a mere play of shadows. ... (84–5)

Thus the phases of the moon are in some sense not real, but merely an appearance.

In response to the second problem, the apparently anachronistic denial of reality to the sensible world, Popper says:

> The great discovery that the Moon is an unchanging spherical body is generalized by Parmenides to the view that perhaps the whole world is unchanging and immovable. Perhaps *all* change, all motion is an illusory play of lights and shadows, a play of light and night? Perhaps one can *prove* that all motion is impossible? (85, Popper's italics)

Thus Parmenides' astronomical discovery leads him to question the reality of change and to hypothesize the changelessness of reality. In a Hegelian way, observation undermines observation and leads the philosopher to choose a changeless reality over a changeable appearance. Thus we get a kind of historical and psychological reconstruction of Parmenides' progress from cosmologist to anti-cosmologist. Indeed Parmenides was "essentially a cosmologist; and so far as he made use of an 'ontological' argument, he used it only as an instrument in an attempt to obtain a cosmological result" (114).

This is an intriguing story that gives pride of place to Parmenides the cosmologist. It brings with it some problems, however. In the first place, the story asks us to focus on the changelessness of the moon throughout the apparent changes of phase, while at the same time ignoring its motion relative to the sun, without which it would not undergo the apparent changes. Is this motion not real? We can ignore it for certain purposes, but the astronomer or cosmologist must recognize that the relative positions of sun and moon are absolutely crucial for the apparent changes that take place. The moral of the story is not necessarily, then, that observations undermine observation itself, but that observations of one (apparent) change may lead us to the recognition of other relevant changes that explain it. Further, the phases of the moon, understood in their new context, show how there is some sort of causal interaction between two heavenly bodies previously thought to be unconnected. Parmenides the cosmologist should have been inspired to become more aware of the complex interactions between plural existents rather than to dismiss change as illusory.

Popper takes his account one step farther. Using Parmenides' discovery of the source of the moon's light, he holds that we can understand the cosmological principles Parmenides develops in the Opinion section of his poem. There he points out an error, a "fall," that leads mortals to seek for cosmological explanations (B8.53–54):

According to Parmenides, as here interpreted, the fall consists in the giving of names to two things—light and night—instead of only one—night, the dark Moon, the heavy dark matter. The forbidden move was to name "light"—a no-thing. This is where "they"—the mortals, the intellectual sinners—"went astray." (72)

On this view it is light rather than night that does not merit a place in the correct ontology. Night is the stand-in for real being, and the original sin of ignorant mortals is to accord to light an equal place as a principle of explanation. Popper sees the possibility of a rich and complex interaction between Parmenides, Xenophanes, and Heraclitus in the development of Parmenides' theory.[37] On his view, indeed, the elderly Xenophanes ends up borrowing the model of a spherical earth from Parmenides.[38]

In terms of cosmological implications, Popper sees that once Parmenides recognizes the sun as the source of the moon's light, he can infer that the moon is spherical from the pattern of shadows (133 n. 63). He also believes that Parmenides infers the spherical shape of the sun by an inductive leap (135). It seems possible to use the same logic to infer the sphericity of the earth (even if this for an ancient was a greater leap, given that the earth was considered *sui generis* and not as a heavenly body). Thus one of Parmenides' important astronomical theses might be deduced from another. I do not, however, see that Popper proposed this final inference.

Overall, Popper sees in Parmenides the work of a great scientific innovator. Whether his astronomical hypotheses inspired his ontological theory, how he viewed his cosmology in relation to this ontology in general, how precisely he was influenced by and influenced his contemporaries—all these questions remain difficult and controversial,

---

37. Popper 1998:135–45.
38. Popper 1998: 42, 45, 139.

and Popper's explanations conjectural. Yet of all Parmenides' commentators, only Popper has paid him due respect as a foundational thinker in cosmology and astronomy. It is Parmenides' foundation on which later cosmologists and astronomers built, and without his work the course of astronomy and science would be very different from what it was.

It would be desirable to put Parmenides' insights into the context of his cosmology as presented in the last part of his poem. Unfortunately, the fragments and testimonies for this section are particularly meager and tend to focus on programmatic statements rather than the details of his theory. Scholars have arrived at no consensus about the nature of his cosmology despite efforts to reconstruct it.[39] In the present state of the evidence, Parmenides' own statements in B14 and B15 seem clearer than testimonies about them and how they fit into his larger project.

3.5 CONCEPTUAL ADVANCES

We have no detailed information about Parmenides' astronomical insights other than the fact that he achieved them. We have a statement of the evidence for heliophotism, namely that the luminous part of the moon is always facing the sun (B15). We have no information about how Parmenides might have come to see this relationship. We have noted that an analogy to the illumination of cloud formations might provide a bridge from the meteorological to the astronomical, in a period when the two sets of phenomena were regarded as continuous. Certainly observation enough was not sufficient to arrive at heliophotism, or other inquirers or cultures would have achieved the insight before Parmenides.

---

39. See, e.g., Pellikann-Engel 1974; Bollack 1990.

As to the identity of the morning star and the evening star, we have a brief report:

> Parmenides puts the morning star [*heōios*] first, which he considers to be the same as the evening star [*hesperos*], in the aether; after it the sun, under which are the stars in the fiery region he calls heaven. (Aetius 2.15.7 = 28A40a)

Some sort of observation had perhaps led to Parmenides' insight that the morning star and the evening star are the same entity. What is needed is to track the relative position of, for instance, the morning star, noticing that it gets closer and closer to the sun until it disappears, after which, a few days later, the evening star appears close to the setting sun and gradually distances itself until it reaches a maximum elongation, after which it gets ever closer to the sun until it disappears, after which the morning star reappears. The two stars are never visible at the same time, and, looked at from the standpoint of the sun, form part of a continuous cycle back and forth to the west or east of the sun. The motion of a single body would account for all the appearances and their ordering. A conceptual leap is required, but it could be made only on the basis of good and sustained observations.

Parmenides' conjecture that the earth is spherical seems to come out of nowhere. Certainly all his predecessors envisaged a flat earth, from Homer and Hesiod to Xenophanes. There is one way he might have inferred the sphericity of the earth: having arrived at the theory of heliophotism, he would see that the moon was spherical, based on the pattern of shadows that moved across its face. If the moon is spherical, he might have reasoned, the other heavenly bodies are likely to be so also. Aristotle in fact made this inference.[40] If the earth

---

40. *On the Heavens* 291b22–23, to be discussed below, ch. 4.1.3.

itself is a heavenly body, it too was likely to be spherical. The last step is bolder than the previous one, simply because there was a tendency to view the earth as unique—and obviously flat to all appearances. It is possible also that Parmenides was led to his insight by another consideration.

> Parmenides and Democritus[41] [say the earth] remains in place because it is equally distant from everything, being in a state of equilibrium since it does not have any reason why it should incline this way rather than that. For this reason it only shakes; it does not change its place. (Aetius 3.15.7 = 28A44)

Anaximander seems to have originated the equilibrium view. But his cylindrical earth is not completely symmetrical—not in three dimensions. Parmenides' spherical earth, by contrast, has radial symmetry in three dimensions and hence is maximally balanced with respect to forces from around it, especially if these consist of circular rings (*stephanai*).[42] Whereas the induction from the moon provides an a posteriori reason for sphericity, the argument from equilibrium would offer a decidedly a priori reason.

Parmenides holds that what-is "is complete/like to the mass of a well-rounded ball [*sphaira*] from every direction" (B8.42–43, cf. 49). This passage suggests to some interpreters that Parmenides' what-is is literally spherical in shape, but it suggests to others that what-is is only figuratively comparable to a ball.[43] If what-is is literally spherical, then it may provide the reality behind the deceptive cosmology presented in the last part of Parmenides' poem, and offer a kind of

---

41. The account given here seems inappropriate for Democritus. See John Philoponus *Physics* 262.8–13; Aetius 3.12.2 = 68A96.
42. Parmenides B12 with Aetius 2.7.1 = 28A37.
43. For the literal reading, see recently Palmer 2009:156–7. For the figurative reading, Mourelatos 2008a:129: what-is is "perspectivally neutral." See also Coxon 1986:214, 217.

paradigm for explaining phenomena.[44] Clearly the sphere is a powerful image for Parmenides, but whether he provides a direct argument for the sphericity of what-is remains a moot point. Most readers would take Parmenides' image as the embodiment of an a priori constraint or at least predilection.[45] The sphere as a regulative idea would have a distinguished future in "centrifocal" dynamics and spherical cosmology and astronomy, as we shall see.[46]

While we can identify possible connections among Parmenides' insights, we cannot say for certain either how he came to them or how he defended them, with the one exception noted above. Nonetheless, he seems to have made conceptual leaps that were destined to have a major impact on the future of astronomy.

CONCLUSION

We began by asking a series of questions about early Greek astronomy:

1. Who discovered the theories in question first?
2. What led him to this discovery?
3. Did the two philosophers (Anaxagoras and Empedocles) have good evidence for the theories?
4. Did the community of philosophers accept the theories?
5. Did they develop the theory on their own, or did they borrow it from another source (Thales, Pythagoras, the Babylonians)?

---

44. In B1.29 Truth is called εὐκυκλής, "well-rounded," in the version of Simplicius (but εὐπειθής, "persuasive" in the versions of Plutarch, Sextus, Clement of Alexandria, and Diogenes Laertius; and εὐφεγγής, "shining," in the version of Proclus).
45. On the other hand, Popper 1998 sees Parmenides' interest in the sphere as deriving from his empirical discovery of the moon's sphericity (85, 122, 136), and his astronomy as strongly influencing his ontology (70–71). But I do not find that he ever attempted to connect the sphericity of the heavenly bodies with the *sphaira* of B8.43. It would, nonetheless, be reasonable for Popper to say that the image of the sphere arose a posteriori.
46. See above, ch. 2, n.8.

We have at present provided at least a partial answer to two of these questions. The remarkable chain of events that began theoretical astronomy as we know probably started with the recognition of heliophotism. This theory, or insight, derives, as far as we can tell, from Parmenides of Elea, who, writing in the early fifth century, saw that the moon's phases could be explained on the basis of the moon's position relative to the sun, supposing that the sun was the moon's source of light—just as, perhaps, it is for clouds. It is plausible to suppose that Parmenides came to this insight by himself, unaided by earlier speculations on the moon, which were unhelpful, or Babylonian data and theories, which were most likely unknown to him, and which did not, in any case, derive the moon's light from the sun. The supposition that he had a Pythagorean informant seems gratuitous.

Thus in answer to question (2): Parmenides paved the way. In partial answer to question (5): Parmenides seems to be original in his contribution to the beginnings of astronomy. As to the further development of the theory of eclipses, there is no record that Parmenides had anything to say about eclipses, even if both his predecessors and his successors did. The students of astronomy and doxographers who canvassed early studies for new theories seem to have found nothing on this topic from Parmenides. We can say in answer to (1) that Parmenides (and not either Anaxagoras or Empedocles) discovered the source of the moon's light; as to the explanation of eclipses, question (1) must remain open, as well as questions (3) and (4). Moreover, we will have to see what role Parmenides' insights played in the further development of early Greek astronomy. What difference does it make to know that the moon gets its light from the sun?

Chapter 4

Empire of the Sun

Implications of Heliophotism, and a New Model

The sun imparts to the moon its brightness.

—Anaxagoras B18

[The moon] gazes into the bright circle of her lord's face.

—Empedocles B47

So far we have seen that the most likely inspiration for the astronomical advances of the fifth century is Parmenides, not Babylonian astronomers, whose important observational and mathematical advances seem to have been unknown to the Greeks of this period. As for Thales, the alleged founder of Greek astronomy, we have seen no reason to think that his astronomical theories, whatever they were, influenced his successors. For if he discovered heliophotism, he did not pass it on. Nor are the doxographers able to cite any clear astronomical theory for Thales. In fact, there is a lack of early evidence to support an attribution to him of heliophotism. Nor is there any sign that Pythagoras provided a scientific cosmology or astronomy. Most of the cosmological ideas attributed to him seem to come from Philolaus, who wrote a couple of generations after Parmenides (on Philolaus, see below, ch. 6.3.3).

As we saw in the last chapter, Parmenides makes an important connection in seeing the moon's light as derived from the sun. Heliophotism makes a causal connection between the phases of the moon and the sun:

the sun's light is reflected from the surface of the moon. Since the moon occupies different positions relative to the sun throughout the lunar month, different portions of its surface are illuminated. The luminous portion always faces the sun, providing prima facie evidence for the theory. By tracking the positions of the moon relative to the sun we can verify the predictions made by the theory as to how the phases correlate with the moon's relative position. In effect, heliophotism turns the question of the moon's phases into a problem of geometry and solves it theoretically in a way that corresponds precisely to empirical observations. No other theory had come close to this level of prediction and explanation, and indeed no previous theory had even tried to correlate the phases of the moon to the moon's position relative to the sun. Heliophotism took a phenomenon that had been treated as merely an accident and accounted for it as a necessary consequence. Thus anyone who was seriously concerned with astronomy should have been duly impressed by the startling new theory of Parmenides, the alleged enemy of empirical science.

We have perhaps made a step forward in identifying Parmenides as the beginning of a new approach to astronomy. Yet there remain major gaps in our understanding of early Greek philosophy in general and early Greek astronomical theory in particular. We lack the background information that is often available for reconstructing developments in modern science. We do not know how the transmission of ideas occurred from Parmenides to his successors or how specifically they reacted, except to the degree that their extant writings or reports of those writings reflect his ideas. Yet there is one way to explore developments that can help fill in the gaps in historical evidence: we can ask what the implications of a given theory are, and see to what extent these implications are reflected in later theories. This is an application of a method stated by Plato: "taking as my hypothesis ... the theory that seemed most compelling, I would consider as true ... whatever agreed with this, and as untrue whatever did not so agree" (*Phaedo* 100a, trans. Grube). In the case of heliophotism the implications are especially rich and suggestive for future elaborations and applications.

4.1 ANTIPHRAXIS AND OTHER THEORETICAL IMPLICATIONS

As we have seen, heliophotism finds immediate empirical support. But what of its theoretical implications? A good scientific theory should not simply account for observable phenomena, but also have consequences apart from them that could be explored and tested independently of the initial set of data. In fact, heliophotism does have a number of implications. Some of these may appear trivial to the reader, but we should remember that the competing theories are those we have already studied in ch. 2, not modern theories. What may seem obvious now did not necessarily seem so in the early fifth century BC.

4.1.1 *The Moon Is Opaque*

If the moon reflects the sun's light, it must be opaque, providing a reflective surface to the rays of the sun. In this way it will be different from the clouds that provide a model for Xenophanes and perhaps inspiration for Parmenides, insofar as clouds are translucent and only partially reflective. The fact that the moon is partially dark during some phases and completely dark during the new moon shows that no appreciable amount of light is passing through its body. Otherwise the moon would be partly illuminated even when it was light from the side or the back, as Plutarch noted:

> Were that true, we should see the moon at the full on the first of the month no less than in the middle of the month, if she does not conceal and obstruct [*antiphrattei*] the sun but because of her subtility lets his light through . . . (*The Face on the Moon* 929b–c, trans. Cherniss)[1]

---

1. Plutarch's acute discussion brings together several Presocratic and later theories for examination.

4.1.2 The Moon Orbits Below the Sun

Similar considerations lead us to see that the moon's orbit must lie below that of the sun. For the new moon occurs when the sun and moon are at conjunction, and if the moon were above (farther away than) the sun, then it would be full at conjunction as well as at opposition. We could get a lunar cycle with two full moons and two half moons, but no new moon or crescent moons. The only configuration that allows for the lunar phases we actually see is that in which the moon lies below the sun relative to earth, or in other words, in which the moon's orbit around the earth is wholly inside the orbit of the sun.

Other early theories of the moon had the moon's orbit being lower than the sun's. For Anaximander, the sun had to be farther from the earth than was the moon because it was brighter, more fiery.[2] For Heraclitus, the moon must be lower because it travels in a less pure region of the heaven.[3] Both these philosophers give physical reasons for the relative altitudes of the sun and moon, but neither gives a purely astronomical one. Heliophotism provides a purely astronomical and empirical basis for putting the moon below the sun.

4.1.3 The Moon Is Spherical.

Heliophotism provides evidence that the moon is spherical, or at least spheroid. The dark portion of the moon is explained by heliophotism as a shadow. But a shadow of the sort we observe passing over the face of the moon is possible only on a spherical or roughly

---

2. Hippolytus *Ref.* 1.6.5 = A11; Aetius 2.20.1, 2.21.1 = A21; Aetius 2.25.1, 2.28.1 = A22; see Bodnár 1988.
3. Diogenes Laertius 9.10 = A1, Aetius 2.28.6 = A12.

spherical body. If, for instance, the moon were a flat disk, then when it was lit from the side it would show briefly a partially illuminated surface, then quickly become full. The shadow would not retreat gradually from circular to gibbous to half to crescent to nothing, while the luminous surface advanced from nothing to crescent to half to gibbous to full circle. This point is seen clearly by Aristotle:

> the evidence of our eyes shows that the moon is spherical. For how else would the moon as it waxes and wanes show for the most part a crescent-shaped or gibbous figure, and only at one moment a half moon? (*On the Heavens* 291b17–21, rev. Oxford trans.)

Strictly speaking, we see only one side of the moon, so that we can affirm only that the moon is a hemisphere; but it seems plausible to think that the moon is symmetrical, and hence spherical.

This point may seem insignificant. But think of preceding theories: according to Anaximander the moon is a vast ring the size of the moon's orbit, and the visible part an aperture in the ring. According to Anaximenes the moon is a flat circle like a leaf. According to Xenophanes the moon is a cloud. And according to Heraclitus, the moon is a bowl-like body, hemispheric in shape, but hollow and full of fire, so that when we look at the full moon we see only the fiery contents of a concave vessel.[4] To have evidence, empirical evidence, that the moon is a spherical body is a major advance.

But if one body is spherical, perhaps all are. For whatever forces or factors work on the moon presumably work on other bodies as

---

4. As we have seen (above, ch. 2.5), on one possible reconstruction, the fire in the bowl may be spherical in shape, in which case the moon will appear spherical, although the solid portion of the moon will be a concave hemisphere, like a bowl.

well. This further implication requires an inductive leap. But it is at least consistent with early theoretical methods and models. For Anaximenes, all heavenly bodies are rings. For Anaximenes at least all free-floating bodies are leaf-like. For Xenophanes all heavenly bodies are clouds or cloud-like. For Heraclitus the sun and moon are bowl-like. Now Parmenides can prove empirically that one heavenly body is spherical. Aristotle himself is prepared to make an inductive generalization from this fact: "One, then, of the heavenly bodies being spherical, clearly the rest will be spherical also" (*On the Heavens* 291b22–23).

At this point we have a possibility of seeing two of Parmenides' insights as connected: if Parmenides fully understood heliophotism, he would see that the moon provides a model for all the heavenly bodies. Now it takes a greater leap to see the earth, which is generally regarded as *sui generis* in early theories, as a heavenly body. Nevertheless, if we apply the same principle to the earth as to the bodies "above" the earth, then the earth itself becomes a spherical body (as Aristotle would later allow). Hence we glimpse the kind of powerful theory-building in Parmenides that will later seem obvious to Aristotle. Heavenly bodies, including the earth, must, by parity of reasoning, be spherical.[5]

4.1.4 The Sun and the Moon Are Permanent Bodies

Anaximander and Anaximenes view the heavenly bodies as permanent fixtures of the heavens. But Xenophanes and Heraclitus have reservations about this. For Xenophanes the reports are ambiguous.

---

5. Karl Popper (1998), who has perhaps put more emphasis on Parmenides' scientific discoveries and their philosophical implications than anyone else, sees the moon's shape as an immediate inference from heliophotism, but curiously he does not connect the earth's shape with the insight (see 70, 122, 133 n. 63).

In one account, the sun comes together at dawn as little flares congregate into a globe.[6] In another account, there are a series of suns. One sun rises in the east, passes over a given area of the earth, and then continues onward to the west; the next day another sun arrives from the east, passes over, and continues on.[7] In the infinite plane of earth there are other suns to the north and south of our position which continue the same east to west procession, so that the multiple suns present the same phenomena, but in different regions.[8] As we have seen (ch. 2.2.4), it might be possible to reconcile the two reports if there is some sort of nucleus that travels continuously, and which serves to attract the combustible vapors from the earth which seem to fuel the sun's fire. During the night conditions are not right for ignition of the sun, but in the morning fiery manifestations gather together into an orb.[9]

For Heraclitus, we are told, "the sun is new every day."[10] The sun and moon consist of a bowl-like structure which is filled with fire arising from earthly vapors.[11] It is possible that the structures remain constant, but the fires are ephemeral as one quantity of fuel is burned up to be replaced by another. We have no detailed cosmography for Heraclitus, so we cannot say what happens to the sun and moon at night. In any case, he seems to stress the discontinuity rather than the continuity of the astronomical bodies.

---

6. Aetius 2.20.3 = A40; Lucretius 5.660–65; Diodorus Siculus 17.7.5–7; Pomponius Mela 1.94–5; Hippolytus 1.14.3 = A33; ps.Plutarch *Miscellanies* 4 = A32; Keyser 1992.
7. Aetius 2.24.4 = A41, 2.24.9 = A41a.
8. Mourelatos 2002a:332–337.
9. Mourelatos 2008b:143–145 gives a reconstruction stressing that what is gathered is moist vapor that is then inflamed by condensation. According to him there is a series of suns that travel from east to west (in one given zone, and another series for each zone); each sun continues until it is eclipsed, after which another sun is gathered from exhalations (138–39). I think it is possible that as a sun approaches our area at dawn it gathers together vapors from nighttime exhalations to strengthen itself.
10. Aristotle *Meteorology* 355a13–15 = B6.
11. Diogenes Laertius 9.10–11 = A1; Aetius P 2.28.6 = A12.

Heliophotism does away with this impermanence, whether of identity (as with the different suns of Xenophanes on one account) or of physical continuity. During the middle part of the lunar month, the moon is visible all light long and illuminated for its whole transit of the sky. Since, by hypothesis, the sun is the cause of the moon's illumination, it follows that the sun must exist for the whole of the night as well. That is, even when the sun is invisible to us from our position on the earth, we can infer its existence from solar light reflected by the moon. Since this happens during part of the lunar month, we seem justified in believing that every night the sun is shining through the night, even when we see neither it nor its reflection from the moon.

Similarly the moon, as an opaque, spherical body, must remain constant in shape and character throughout the lunar month. For all but perhaps three days of the month it is visible by virtue of the sun's reflection of its surface. While the phases change, meaning the appearance of the illuminated portion varies, we can predict and explain that variation with reference to the constancy of the moon's shape and the variability of its position relative to the sun. We must, accordingly, suppose that the moon continues to exist during the new moon, when the sun's light is reflecting off the far side of the moon, where it is facing the sun in the latter's higher orbit. The theory tells us that the moon is no less present when we cannot see it than when we can, and it also tells us why we cannot see it during the end of the lunar month.

At first blush the point that the sun and moon are permanent bodies appears trivial. But it rules out important competing theories about the nature of the large heavenly bodies. Furthermore, to the extent that heliophotism is supported by empirical evidence, we can say we have empirical evidence for the continuity of the sun and moon. Theories that make them ephemeral manifestations are obsolete and indefensible.

4.1.5 *The Heavenly Bodies are Massy*

If the moon is opaque, spherical, and permanent (implications (1), (3), and (4)), it must be a three-dimensional solid body. If so, it must be earthlike in its composition and accordingly solid, massy, heavy. And if we extend this insight by induction, the other heavenly bodies must be likewise solid, massy bodies. The view that they are composed of fire or air, or consist of a thin leaf-like structure must be false. If this is so, some new physics or astrophysics is needed to account for their existence and motion—to explain how large, heavy bodies can stay in orbit around the earth.

4.1.6 *The Paths of Some Heavenly Bodies Go Under the Earth*

We have seen several competing and incompatible accounts of the paths heavenly bodies follow. According to Anaximenes, they always remain above the earth. According to Thales, the earth floats on a vast sea, which seems to block the passage of the heavenly bodies under the earth. According to Xenophanes, the sun and moon travel straight onward without end, from east to west, while the plane of the earth's surface extends to infinity in every direction. (Apparently the stars turn around a center above a given point on the earth.) Heliophotism has important implications for orbits. In Anaximander the rings of the heavenly bodies go below the plane of the earth, and may possibly pass under the earth, though this last possibility is not clearly depicted.

On the basis of heliophotism we know that the moon is closer to the earth than the sun. We know that the moon shines all night during the middle of the month; hence the sun is shining all night long, at least in the middle of the month. If this is so, it is plausible to think that the sun is shining all night long throughout the month. But we also know the relative positions of sun and moon remain roughly constant from day to day, gradually changing a few degrees as the moon proceeds in an eastward direction relative to the sun. During the full moon, the sun and moon are in opposition, approximately 180° from each other. During the night of the full moon, then, we can track the position of the sun throughout the night even though we cannot see the sun. By parity of reasoning we can track the moon throughout the day when it is invisible to us.

We can establish, on the basis of a well-confirmed theory and empirical evidence, then, that the sun is traveling *around the earth*. Furthermore, since the moon at night is above the earth and the sun is in opposition, it must be *below the earth's plane*. Accordingly, any theory which holds that the earth is an infinite plane, such as was apparently the view of Xenophanes, or that the sea is vast or infinite in extent, as may be the view of Thales, is false. Furthermore, any theory which holds that the heavenly bodies are confined to the space above the earth, as was the view of Anaximenes, must also be false. The paths of the sun and moon go below the earth, which then must in some way be suspended in open space, in the manner suggested by Anaximander's model. And only a theory that allows for the free motion of heavenly bodies under the earth will provide an adequate physical basis for astronomy.

Let us pause here for a moment and notice what heliophotism does for early (pre-Parmenidean) astronomical theories. It refutes them one and all. The view that the earth or sea is vast or infinite in extent is incompatible with heliophotism. The view that the sun and moon are ephemeral manifestations is untenable. The related view that heavenly phenomena are continuous with meteorological phenomena, that perhaps the ignition of their fires is dependent upon vapors rising from the surface of the earth is exploded. The view that the shape of the heavenly bodies is a vast ring, or a disk, or a flat leaf, or a cloud-like texture, are all demolished. To the extent we find heliophotism to be supported by empirical observations, we can say that there is empirical evidence to reject all of these theoretical features. To the extent we find heliophotism compelling, we gain irrefutable proof that the structure and path of the heavenly bodies is misrepresented by early astronomical theories. Even one feature on which virtually all theories agree, namely that the sun is higher than the moon, is newly supported on the basis of empirical evidence drawn from astronomical observations rather than on the basis of a priori theorizing. In this case there is an agreement on the relative position of the sun, but a very different appeal to evidence. In every way heliophotism undermines the early theories and establishes its own preeminence.

4.1.7 Eclipses Can Be Explained by Astronomical Alignments

Ultimately, however, the most powerful evidence for heliophotism will be its ability to account for eclipses, both solar and lunar. As I have suggested already, heliophotism does not entail the correct account of eclipses. It could be that some other mechanism is responsible for the darkening of the sun in a solar eclipse or the darkening of the moon in a lunar eclipse. But as we shall see, it does suggest the correct account, while the correct account presupposes heliophotism.

As we have seen, heliophotism entails that the moon is (1) an opaque, (3) spherical body (2) orbiting below the sun; that (4) both it and the sun are permanent (5) massy pieces in the furniture of the heavens, (6) whose orbits carry them below the earth. Heliophotism itself tells us that the source of the moon's light is the sun. If the moon lost its light, an obvious possibility for this would be some interference with the sun's light. Now the model of the heavens suggested by heliophotism presents one significant possibility for an interference with the sun's light. During the middle of the lunar month, when the moon is full, the sun and the moon are, according to the geometry of the model, in opposition, and the earth stands between them. On all early models of the heavens, the earth is relatively large in relation to the sun and moon. Thus we can say that the model allows and perhaps even predicts the possibility of a blocking of the sun's light by the earth, which would perforce darken the moon. The possibility of this kind of interruption of the sun's light is built into the model. The Greek word for blocking or interposition is *antiphraxis*. Thus to picture the model entailed by heliophotism is to conceive the possibility of a lunar eclipse being caused by antiphraxis.

The model derived from heliophotism provides another possibility for solar eclipses. According to heliophotism, the moon is an opaque body below the sun. When it is not illuminated by the sun, it is dark and invisible (as parts of it are when other parts are visible). The theory itself tells us that the moon is wholly invisible precisely when the luminous face is on the backside of the moon; that is, when it is in conjunction with the sun. Indeed, the geometry of the model entails that during the beginning of the lunar month the dark moon is always lurking in the vicinity of the sun. Further, we can observe at various times that the sun and the moon are roughly the same apparent size in the heavens. Thus if the two bodies were in complete conjunction, the moon, being below the sun, would intervene between the sun and earth, and being opaque, would screen or block or intercept (*antiphrattein*) it. This hypothetical event would be another case of antiphraxis, with the moon blocking the sun's light to earth.

Thus the model produced from the implications of heliophotism points to antiphraxis as a possible explanation of eclipses, both lunar and solar. For its part, antiphraxis only makes sense on the basis of heliophotism: the sun must be the only major source of light in the heaven, and the moon a dark body with a reflective surface, for antiphraxis to work. As we saw earlier (above, ch. 3.2), antiphraxis presupposes heliophotism and could be developed only as an application of heliophotism.

The beauty of this relationship is that, given heliophotism, no other assumptions are needed to account for eclipses than the sun, moon, and earth themselves. Eclipses can in principle be explained on the basis of geometrical alignments: when the sun, the earth, and the moon line up precisely in a "syzygy," the earth will block the sun's light to the moon, the moon will lose its light, and the moon will be darkened. When the earth, the moon, and the sun are in perfect alignment, the moon will block the sun's light to earth and the sun will be darkened. Eclipses are a matter of celestial geometry, not of stopped up apertures, revolving bowls, high-altitude clouds, or failures of fuel, much less of occult properties.

If this account is correct, moreover, eclipses can occur only at an appropriate point in the orbits of the bodies. A lunar eclipse can happen only during the time of a full moon, when sun and moon are in opposition, that is, on opposite sides of the earth. A solar eclipse can happen only at the time of a new moon, when sun and moon are in conjunction. Notice that no other early theory of eclipses entails any such stringent requirements. As far as we can tell, all other theories allow eclipses to happen at any point in the lunar month. In this sense, antiphraxis admits of conditions of falsification that other theories do not. As Karl Popper has stressed, falsifiability is a sign of a theory's scientific content.[12] The theory allows us in principle to refute it with the next eclipse. So every eclipse becomes a test of the antiphraxis theory.

So far the discussion has centered only on theoretical implications, in relation to implications of other competing theories. At this point we can observe that heliophotism is extraordinarily rich in its implications and suggestive for future applications.

---

12. Popper 1968. Popper discusses Anaxagoras as the discoverer of the true theory of eclipses (Popper 1998:133n.63), but does not bring up the topic of falsifiability in this context.

4.2 A NEW PHYSICS

Heliophotism is clearly incompatible with the Meteorological Model of astronomy that prevailed before Anaxagoras (ch. 2.3 above). To the degree that it finds empirical support, it undermines the conception of astronomical events as continuous with meteorological or atmospheric phenomena. According to the Meteorological Model, heavenly bodies are lighter than air, or at least light enough to be supported by winds and movements of air, like a kite or a blown leaf. But if we suppose that at least one large heavenly object is spherical and opaque, we raise the question of the physics of heavenly bodies. The moon seems to be solid and, as we shall see, "earthy" rather than airy—for it even looks earthy (rather than, e.g., cloudlike) to critics of the Meteorological Model. If that is so, winds and breezes will not account for the revolution of the heavenly bodies. Some stronger force is needed to keep heavy bodies aloft.

To use distinctions that Aristotle emphasized later, if the heavenly bodies are earthy or rocklike or otherwise solid and massive, they will not stay above the earth by natural lightness. They will be there "contrary to nature," and will need some sort of violent force to keep them airborne. If by some chance they lose the support this force gives them, they could fall down to earth. The model suggested by the astronomy of solid bodies will require a new physics with powerful forces to counteract the natural tendencies of heavy bodies to fall.

4.3 ANAXAGORAS' NEW COSMOLOGY AND ASTRONOMY

Thus far we have dealt with the implications of heliophotism for astronomy and ultimately cosmology. This has been a largely a priori exercise based on the hypothesis of heliophotism. Let us now turn to the theory of Anaxagoras. We have a detailed report of his physical theory that is interesting precisely because of the connections it makes with the implications of Parmenides' insight.[13]

> [3, 4, 5] The sun and moon and all the heavenly bodies are fiery stones carried around by the revolution of the aether. And there are below the stars certain bodies invisible to us which are carried around with the sun and moon. We do not feel the heat of the stars because of their great distance from the earth; moreover, they are not as hot as the sun because they occupy a colder region. [2] The moon is below the sun and nearer to us. The sun exceeds the Peloponnesus in size. [H] The moon does not have its own light, but gets it from the sun. [6] The revolution of the stars carries them under the earth. [7 = A] The moon is eclipsed when the earth blocks it, or sometimes one of the bodies below the moon; the sun is eclipsed when the moon blocks it at the time of the new moon. The sun and the moon make their turnings [i.e., solstices] when they are deflected by the air. The moon makes frequent turnings because it cannot overcome the cold. He [Anaxagoras] first correctly explained eclipses and illuminations. [1] He said the moon was earthy and had in it plains, <mountains>[14] and valleys. (Hippolytus *Refutation of All Heresies* 1.8.6–10 = 59A42)

In the brackets I have tagged the principles of heliophotism (H) and antiphraxis (A), along with instances of the seven implications of heliophotism noted above.

---

13. Fehling 1985:226–229 has argued that Parmenides wrote after Anaxagoras, thus reversing the order of influence; but his arguments for revising the chronology seem to me inadequate.
14. Supplemented by Marcovich from Aetius 2.25.9.

(1) The moon is earthy and hence opaque. (2) It lies below the sun.[15] (3, 5) The sun, moon, and stars are like fiery stones, hence solid, massive bodies of presumably roughly spherical shape;[16] (4) they are permanent bodies rather than ephemeral formations. (6) They, or at least some of the heavenly bodies, are carried below the earth on their orbits. (7) Lunar and solar eclipses come about by the blocking of the sun's light. Anaxagoras' account of eclipses as reported here is actually more complicated than that suggested by heliophotism alone: he allows not only the earth but certain "bodies below the moon" to block the sun's light. In fact he holds that there are other dark bodies besides the moon which can potentially interrupt the sun's light. I shall call these by their modern name, asteroids.[17] As we shall see, Anaxagoras' account of the moon's light is also more complicated than we would expect.

But for now, it is important to notice that Anaxagoras seems to grasp all the implications of heliophotism, besides embracing that key insight himself. Indeed, we have his recognition of heliophotism in his own words, as recorded by Plutarch:

> The sun imparts to the moon its brightness. (*The Face on the Moon* 929b = B18)

Anaxagoras seems to have constructed his astronomical theory on the basis of an appreciation of heliophotism. Indeed, Hippolytus

---

15. Curiously, Gershenson and Greenberg 1964b:47–48 say, "we do not know if he [Anaxagoras] made this assumption nor have we any reliable reports from which we could conclude that he constructed a theory of lunar phases or eclipses."
16. Popper 1998:133n.63 suggests that Parmenides B10 already portrays the moon as spherical. But the term *kuklōps* in line 4 implies only that the moon is circular, not spherical; Parmenides may hold the latter interpretation, but there is no clear evidence for his doing so. On the term, cf. Hesiod *Theogony* 144–45.
17. The position of Anaxagoras' asteroids is of course different from that in contemporary astronomy: below the moon rather than (predominantly) between the orbits of Mars and Jupiter.

reports, "He first correctly explained eclipses and illuminations." This notice, which may go back to Theophrastus and Eudemus, who were concerned to establish the first discoverer of various explanations, is historically plausible in light of the information we have seen.[18] It is, however, debatable in light of the roughly contemporaneous contributions of Empedocles; we shall look into the question of priority in the next chapter.

We may observe that nothing in Hippolytus' account says specifically that Anaxagoras' theory was actually derived from heliophotism. This could be taken as an argument against the claims I have been making that heliophotism is the key to further elaborations in astronomical theory. But we need to recognize that the kind of information Hippolytus drew on came ultimately from doxographical accounts that enumerated the doctrines of individual philosophers taken out of context and juxtaposed with those of other philosophers for the purposes of comparison and contrast.[19] At some point these *disiecta membra* were extracted from this kind of dictionary of opinions and reassembled around particular philosophers, for example Anaxagoras and Empedocles.[20] Hence the list of opinions of Anaxagoras represents a reassembly of disconnected parts by some later scholar who may have had no access to Anaxagoras' original work. If, then, Hippolytus makes no distinction between premises and conclusions in this account, it well may be that whatever logical and theoretical connections Anaxagoras made originally were lost in transmission. In any case, it is not clear how careful early philosophers were in articulating the reasoning behind their theories. They may have presented the finished theory without much comment on

---

18. Eudemus fr. 145 Wehrli = 13A16, Panchenko 2002:324–326.
19. For an introductory discussion, see Mansfeld 1999:28–32; Runia 2008:35–37. For an argument for the interpretation, see Mansfeld 1992a.
20. Mansfeld 1999: 36. On Hippolytus as a source, see Mansfeld 1992b.

how they came to it or on what evidence the theory was justified. Accordingly, we are forced to reconstruct theories and sequences of ideas without the benefit of biographical details, conceptual frameworks, or logical connections.

Since Anaxagoras has heavy, earthy or stony bodies moving above the earth, he needs a physics that can account for this. We are told that he has just such a theory:

> Anaxagoras said that the whole heaven was composed of stones. They are held together by the violent vortex [*sphodra peridinēsis*], and if they escape it they fall down. (Diogenes Laertius 2.12 = 59A1)

Hippolytus says (in the passage quoted earlier) that "the heavenly bodies are fiery stones carried around by the revolution of the aether" (1.8.6). In Anaxagoras' own words we can read of the power of the vortex:

> In this way these things were revolving and being separated off by force and speed (for speed generates force). The speed of them is like the speed of no object of those currently found among men. But it is altogether many times as fast. (B9)

Anaxagoras tells us that Mind began the rotation, which is gradually expanding to take in more and more matter, and producing separation among the elements and articulating a cosmos (B12). The important point for our study is the fact that Anaxagoras supplies the powerful forces needed to hold massy bodies aloft over the earth. Winds or meteorological principles will not suffice, as Anaxagoras recognizes. He must have something like a centrifugal force keeping the heavenly bodies from crashing to the earth. On such a theory it is, in principle, possible for a heavy body to fall out of the sky to earth. And about this we shall have much to say in the next chapter.

Before Anaxagoras some philosophers may have posited a vortex. Anaximander has a circular motion generating the cosmos, and it may continue as the force causing the heavenly rings to revolve.[21] There is some evidence for a vortex in Anaximenes, strengthened by a new testimony:

> They [earlier cosmologists] construct walls in a circle [around the earth] so that they may screen us against the vortex [*dina*], as it whirls around on the outside of the earth, for all those who drive the heavenly bodies around in a circle overhead. (Herculaneum Papyri 1042.8.vi., Epicurus *On Nature* IĀ [33] Arrighetti)

Here Epicurus criticizes his predecessors. Since only Anaximenes constructs mountains around the earth (specifically, to the north) and has the stars circle overhead, the most obvious target here is Anaximenes.[22] Nonetheless, Epicurus may be reading later dynamic theories back into an earlier era, when something less than a powerful vortex was needed to push the stars.

Xenophanes seems to require one sort of high-altitude wind or jetstream to carry the suns and moons from east to west, and some sort of circular whirlwind to move the stars in a circular pattern over our heads.[23] There is a long-standing debate among scholars as to whether the early Ionians posited a vortex motion in the heavens.[24] What one can say is that there is no clear *need* for a vortex theory in

---

21. Hippolytus *Refutation* 1.6.1–2 = A11; ps.Plutarch *Miscellanies* 2 = A10; neither is explicit on the mechanics of cosmic motion.
22. See Perilli 1992.
23. Mourelatos 2002a:347.
24. Zeller 1919–1920:298 n.4 argues against a vortex for the earliest philosophers; Heidel 1906 argues for an early vortex theory. In recent times Ferguson 1971 has argued against an early vortex theory and Perilli 1992 for it. I am inclined to agree with the recent assessments of Sider 2005:134 that "the originator of the vortex theory is Anaxagoras"; and Gregory 2007:116 that "Anaxagoras was the first to employ a vortex in cosmogony."

early Ionian theories, which do not put heavy bodies in orbit. Whatever precisely the forces were that drove the heavenly bodies in earlier theories, Anaxagoras clearly posits a powerful vortex, as does his contemporary Empedocles.[25]

We noted by the way that Anaxagoras' astronomical theory is actually more complicated than that strictly required by the implications of heliophotism. Let us look briefly at the complications and the reasons Anaxagoras might have had for introducing them. One complication is that of having asteroids below the moon. Now these are, according to Anaxagoras, dark bodies invisible to sight, so that by hypothesis direct observational evidence for them is ruled out. Why then does Anaxagoras need them? The most likely reason is to provide for lunar eclipses that are difficult to explain by reference to the earth's blocking the sun's light to the moon. Anaxagoras holds that the earth is flat. Thus if the moon were seen in eclipse when it was rising or setting, it would seem impossible for the earth to be the cause. The earth would then present only a thin straight line to the sun and the moon, while the moon was completely eclipsed. There would, then, have to be some other body blocking the sun's light to the moon—assuming that antiphraxis provides the only plausible account of eclipses. Of course, if the earth were spherical, the moon could be eclipsed on rising or setting; but Anaxagoras had empirical evidence, as he thought, for the flatness of the earth:

> Some think [the earth] is spherical, some flat and drum-like in shape; they find evidence for this view in the fact that when the sun sets or rises the earth makes a straight line where it covers the sun rather than a curved one, as it should, if the earth were spherical, appear in its segment. (Aristotle *On the Heavens* 293b33–294a4)

[25]. See Tigner 1979; Tigner 1974.

> The shape of the whole earth is not flat, as they who compare it to the figure of an extended disk believe, nor concave as others believe, who have said that rain descended into the lap of Earth,[26] but Dicaearchus maintained that it is round, that is, spherical. . . . Whereas the latter view is more disreputable insofar as it is vitiated by an untenable assumption, it is worthwhile to consider the former, which even the natural philosopher Anaxagoras subscribed to, inasmuch as he is said to have contributed certain arguments in favor of it. For he said that the flatness of the earth was clearly proved by the rising and setting of the sun and moon, which, as soon as the brightness of their first light appears, are immediately brought to our view in straight lines. (Martianus Capella 6.590, 592)

Aristotle provides a sketch of an argument by philosophers he does not name. Martianus Capella informs us that Anaxagoras devised the argument, providing evidence for the flatness of the earth.[27] His point is that the earth seems to make a straight line across the rising or setting sun or moon, rather than a convex curve that one would expect from a spherical earth. (See figure 4.1.)

Now if the earth is a thin, flat disk (as Anaxagoras held) rather than a sphere, problems can arise for lunar eclipses. A lunar eclipse that happens either around sunrise or around sunset (and these do

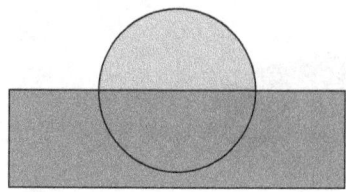 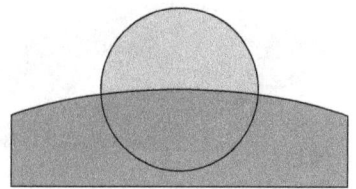

Figure 4.1a. Setting Sun, Flat Earth.   b. Setting Sun, Spherical Earth.

26. Cp. Virgil *Georgics* 2.325–26.
27. See Panchenko 1997.

occur), what I shall call a *crepuscular eclipse*, cannot be explained by the interposition of the earth. For the earth would offer only a thin, straight line in profile to the sun. (See figure 4.2.) In fact the shadow that falls on the moon in a lunar eclipse is always circular in shape. Thus an eclipse of the moon observed when the moon is either rising or setting needs another explanation than the interposition of the earth. If we stick to antiphraxis as the account of all eclipses, we need another body to perform the work of blocking the sun's light. This is where another dark body in orbit could intervene between the sun and the moon. An asteroid must be blocking the sun's light.

Anaxagoras' physics allows for heavy bodies to be carried around the earth. It is not in principle more difficult to account for asteroids than to account for the sun, moon, and stars as populating the heavens. Asteroids seemed to Anaxagoras to provide the simplest explanation, consistent with the theory of antiphraxis, for lunar eclipses in which the earth's broad, flat surface was not interposed between the sun and the moon.

There is another complication in Anaxagoras' astronomy, in his account of lunar light. While he has a perfectly good and empirically verifiable account of the phases of the moon as reflections of solar light, he nevertheless allows another source of lunar light as well.

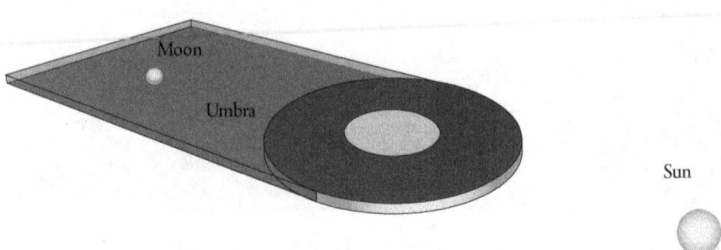

Figure 4.2. According to Anaxagoras' cosmography, a crepuscular eclipse of the moon should produce a shadow across the moon in a straight line. But in lunar eclipses only circular shadows are observed.

> The followers of Anaxagoras and Democritus say the Milky Way is the light of certain stars. When the sun travels under the earth it does not look on some of the stars. The light of any that it looks upon does not appear because it is prevented by the rays of the sun. But for all those which are blocked by the earth [*antiphrattei hē gē*] so that the sun does not look upon them, their natural light, they say, makes the Milky Way. (Aristotle *Meteorology* 345a25–31 = 59A80)

> A third view is that of Anaxagoras and Democritus. They say the Milky Way is the proper light of stars not illuminated by the sun. For the stars, he says, have their own light as well as a light acquired from the sun. And the case of the moon makes this clear. For this has one kind of light of its own and another from the sun. Its own light is coal-like, which the moon's eclipse shows us. (Olympiodorus *Meteorology* 67.32–37)

Aristotle distinguishes between the natural light of certain stars and reflected light. In explicating this Olympiodorus makes it clear that the moon, too, has a natural light distinct from the reflected light of the sun.

The commentator offers as evidence for the natural light of these heavenly bodies the "coal-like" glow of the moon during an eclipse. In fact during many lunar eclipses, the moon takes on a reddish hue while it is in the umbra of the earth. The appearance of the moon could be compared to the glow of a coal from a dying fire. Anaxagoras seems to have observed lunar eclipses carefully and to have wanted to account for the light that is seen emanating from the moon even during its complete eclipse. The moon must have a natural source of light that is normally overpowered by its reflection of the sun's light. But when it is deprived of the sun's light, we see the natural glow of the moon.[28]

It would be helpful to know more about the moon's own light and its cause. One could suppose that it is a residual light from the sun's

---

28. See Dreyer 1953:32 n.1; O'Brien 1968:125–127; Panchenko 2002:328–330.

illumination: the sun might heat up the moon's surface, causing it to glow until it can radiate the excess heat. This picture is suggested by Plato:

> This light around the moon is always new and old [*neos, henos*], if the followers of Anaxagoras are right. For as the sun is always traveling around the moon in a circle, presumably it is always casting on it new light, while that remaining from the previous month is old. (*Cratylus* 409b = 59A76)

Playing with a name for the first day of the month, Plato's Socrates suggests that there is a residual light from the old month present in the moon. On the other hand, Aristotle's mention of a "natural" light suggests that this light is not a left-over from an earlier illumination, but a feature that the moon would have with or without the sun's action. It is possible also that this light is a result of friction or some other physical interaction independent of the sun (although this is not what Aristotle normally means by a natural manifestation).

It is possible he also used this secondary light to account for "earthshine," the dim light that makes the dark side of the moon barely visible during a crescent moon. (In fact earthshine results from light reflected back to the moon from the earth—in other words, the sun's light falls on the earth's surface, is reflected back to the moon, and then back to the earth.) We find, however, no express discussion of earthshine in ancient reports of early astronomical theories.[29] Yet it may be that Plato's discussion of old light might be a case of "the Old Moon in the New Moon's arms," that is, dim light embraced by the new crescent.[30]

In any case, the stars of the Milky Way, and perhaps all stars, allegedly share in this natural light. But the sun's light reflecting off these (all?) stars outshines their natural light. This action does not seem to

---

29. Conche 1999:236 thinks that Parmenides must have attributed an innate source of light to the moon to account for earthshine, but without any evidence.
30. See Couprie 2011:177.

affect the asteroids, which are always invisible, but it does potentially affect the other heavenly bodies. But even if the asteroids do not glow, why are they never illuminated by the sun so that they reflect it as does the moon and as do the stars? Perhaps they are so close to the earth that during the day any illumination falls on the far side, away from the earth, and during the night they move in the shadow of the earth. This might allow for some bodies higher up to reflect the sun's rays. Could Venus be such a body? Unfortunately, we get no reports on the planet from our sources. But given Parmenides' recognition of Venus, we could expect Anaxagoras to have some theory about it. In fact, of course, Venus is illuminated by the sun. Democritus, whose astronomy is heavily influenced by that of Anaxagoras (as in his view of the Milky Way cited above) puts Venus between the sun and the moon.[31]

We have not discussed all aspects of Anaxagoras' astronomy, but so far what we have seen seems self-consistent. The heavenly bodies are stony or earthy bodies held aloft by a powerful vortex. The sun is the source of most of the light in the heavens. It illuminates the surface of the earth, the moon, and the stars (some or all), and perhaps also the planets. The moon and some or all stars have a secondary light independent of, or perhaps residual from, the sun, which in most cases is obscured by the bright rays of reflected solar light. There is one general theory of eclipses: antiphraxis. Solar eclipses are caused by the occultation of the sun by the moon. Lunar eclipses are caused either by the earth's blocking the sun's light to the moon, or by certain asteroids' blocking the sun's light to it. This theory is more complicated than it needs to be: there is in fact no need to suppose the moon has its own light: during a lunar eclipse the reddish glow results from refraction of the sun's rays through the earth's atmosphere, at the rim of the earth as seen from the moon; during a crescent moon the slight illumination of the dark part of the moon results from earthshine,

---

31. Aetius 2.15.3 = 68A86.

solar light reflected from the earth back to the moon. The earth is spherical and hence never presents a thin line to the sun. Thus the earth's imposition can explain all lunar eclipses without an auxiliary hypothesis. Nevertheless, given Anaxagoras' knowledge and assumptions, we can see why he thought he needed to complicate his theory. His complications were not arbitrary or gratuitous, but represent rational attempts to account for phenomena that Anaxagoras could not account for on the basis of his initial theory.

Most important, we see that a whole range of innovations to astronomical theory introduced by Anaxagoras follow from Parmenides' insight of heliophotism: (1) the moon is opaque; (2) it lies below the sun; (3) it is spherical; (4) the sun and moon are permanent bodies; (5) the heavenly bodies are massy; (6) their paths go beneath the earth; and (7) eclipses, both solar and lunar, are caused by blocking (*antiphraxis*) of the sun's light. The recognition of antiphraxis, requiring as it does a conceptual advance from heliophotism, marks a major step forward in theoretical astronomy. By the time of Anaxagoras, the implications of heliophotism have been explored and accepted. Heliophotism has vindicated itself both by reference to the phases of the moon and to eclipses. And the sun has emerged as the lord of the heavens in a more profound sense than the god Helios had ever been.

4.4 THE LITHIC MODEL

Thus far we have explored the theoretical implications of heliophotism. We have found that they include antiphraxis and several other consequences that constrain cosmography and suggest a coherent model of the heavens that is radically different from that of the sixth century. We have found that all of these consequences are accepted by Anaxagoras and incorporated in his cosmology. Minimally, this study suggests that Anaxagoras engaged in a rational process of theory construction in which he explored and ultimately exploited the implications of Parmenides' great insight. At one level, Anaxagoras was just philosophizing—following out a hypothesis—as any good philosopher should do. But because the initial insight had an explicit empirical basis, he was also consciously developing a theory that had implications for empirical testing and confirmation, for instance in the study of eclipses. And that fact suggests that he was doing something that no Greek philosopher had done before: articulating a scientific theory.

A new model of the heavens emerges in Anaxagoras, one very different from the Meteorological Model of the sixth century with its vapors, exhalations, and clouds supplying fuel and motion to heavenly conflagrations. In Anaxagoras we find earthy or stony bodies torn away from the surface of the earth and hurled violently in a turbine motion. The motion somehow generates heat, which is not that of the combustion of fuel. Besides luminous heavenly bodies there are dark bodies, invisible except when they collide or interpose themselves or fall out of orbit. Because of the central role of rocky bodies in this scenario, I shall call the new picture the Lithic Model. As we shall see, it will become the dominant model of the fifth century, replacing the Meteorological Model, providing new explanations, and benefiting from new empirical evidence. Crucially, the new model makes hardly any connections between meteorological phenomena and heavenly bodies. It severs the heavenly bodies from earthly processes except insofar as it may see boulders as being torn from the earth by a vortex motion and launched into orbit, where the vortex itself is cosmic rather than meteorological in its character. The new cosmology and astronomy are increasingly disjoined from meteorology and near-earth phenomena.[32]

---

32. Already in Parmenides' cosmology there is a distinction between heaven (*ouranos*) and the vicinity of the earth (*ta perigeia*): Aetius 2.7.1 = 28A37.

Before we look at other cosmological theories of the fifth century, we must raise what may seem the relatively pedestrian question of priority: did Anaxagoras himself first draw out the implications of heliophotism, or did Empedocles, Anaxagoras' slightly younger contemporary, who held roughly the same views of astronomy, first make the connection? If the question is pedestrian, the answer is anything but routine. Historical and astronomical events surrounding the development of Greek astronomy in the first half of the fifth century provide some important clues concerning the genesis of the new model, the use of observation and evidence in relation to hypothesis, and ultimately the very acceptance of the model.

Chapter 5

Darkened Suns and Falling Stars

Heaven-Sent Proofs

[Athena hastened earthward] like a star the son of devious Cronus hurled down as a portent to sailors or to a broad host of men, blazing; and from it many sparks fly out.

—Homer *Iliad* 4.75–77

We have seen that the insight of Parmenides served as the impetus for a new understanding of the heavens. Clearly Anaxagoras grasped the implications of Parmenides' insight, and he went beyond it in his understanding of astronomy: he explained eclipses correctly. But his slightly younger contemporary, Empedocles, also understood eclipses. Who then was the first discoverer? Perhaps it does not matter, as long as we can recognize that the generation after Parmenides saw new and important ways to apply his insight to the heavens. But if we can answer this question, we can get a firmer grasp on the development of astronomical ideas in the early fifth century. The question is not an easy one to answer, as we shall see.

There are three possibilities here: Anaxagoras came first, and taught Empedocles, presumably through his writings, since the two philosophers came from opposite ends of the Greek world, and no tradition has them meeting (although biographical inventions establishing links between philosophers seem to have been popular). Or perhaps Empedocles wrote first, and taught Anaxagoras. Or perhaps the two philosophers discovered the nature of eclipses independently, drawing on the same assumptions.

A few basic texts will set the stage. About the relative ages of the two philosophers, Theophrastus says:

> Empedocles ... was born not much after Anaxagoras. (Simplicius *Physics* 25.19–20 = Theophrastus fr. 227A Fortenbaugh = 31A7)

About their relative priorities, Aristotle says:

> Anaxagoras of Clazomenae, being prior to him [Empedocles] in age, but posterior in works, says the principles are infinite. (*Metaphysics* 984a11–13 = 59A43)

Plutarch assigns priority in astronomical theory to Anaxagoras:

> Anaxagoras first put in writing in the clearest and boldest terms of all a theory concerning the illumination and shadow of the moon. This theory, which was not old or generally accepted, at this time still went about whispered in secret with caution rather than confidence, among a few men. For the people did not put up with natural philosophers or those who were called star-gazers, who attributed divine events to irrational agencies, indifferent powers, and necessary effects. But Protagoras was banished and Anaxagoras imprisoned, and was barely saved by Pericles. (*Nicias* 23.2–3 = 59A18)

Hippolytus agrees:

> He [Anaxagoras] first correctly explained eclipses and illuminations. (*Refutations* 1.8.10 = 59A42)

DARKENED SUNS AND FALLING STARS

There is a problem at the outset here. According to Aristotle, Empedocles seems to have written first; according to Plutarch and Hippolytus, Anaxagoras made the first discovery.

In fact it is controversial exactly what Aristotle means to say. On the most straightforward and (some would say) most natural reading of his sentence,[1] Empedocles wrote first although he was younger in age. Yet some scholars think that "posterior" expresses a value judgment on Anaxagoras' works, meaning "inferior"[2]—while others suggest the opposite, "more advanced"[3]—so that, whether Aristotle is slighting or praising Anaxagoras, he is not setting out a chronological ordering of his works in relation to those of Empedocles. Furthermore, Plutarch does not say simply that Anaxagoras came first absolutely, but rather than he was the first to lay out the theory *clearly*, perhaps unambiguously. Since Empedocles wrote in verse and, as we can judge from his many fragments, did not always express himself perspicuously, Plutarch's statement could very well be compatible with Empedocles' having published his writings first.

It appears then that the sources who talk about priority do not help us as much as we might wish. We shall be forced to look at some details of the biographies of the two philosophers, along with some events of the fifth century BC to see which is likely to be the discoverer of the antiphraxis theory. As we shall see, there are many disputed points about the biographies of the philosophers.[4] It may be that astronomical considerations can actually shed light on biographical ones.

---

1. Barnes 1982:306; recently, Curd 2007:133 n.15.
2. Alexander *Metaphysics* 28.1–10; cf. Theophrastus from Simplicius *Physics* 26.7–10, supported by Ross 1924:132 and by Kahn 1960:163–65, who criticizes Ross's translation ("Anaxagoras ... was later in his philosophical activity," which is not in accord with his comment). See also Sider 2005:6–7.
3. Bonitz 1848–1849, 2:67: see Ross 1924 (previous note).
4. On the general question of the priority of writing, see Zeller 1919–1920:1261–64; O'Brien 1968a; Mansfeld 1979, 1980.

5.1 LIVES OF THE EMINENT PHILOSOPHERS

Diogenes Laertius gives us in a nutshell the outline of Anaxagoras' life:

> It is said that at the time of Xerxes' invasion [480] he was twenty years old and he lived seventy-two years. Apollodorus says in his *Chronicles* that he was born in the seventieth Olympiad [500–497] and he died in the first year of the eighty-eighth [428]. He began to philosophize in Athens at the time Callia<de>s was archon [480; or Callias: 456] when he was twenty years old, as Demetrius of Phaleron says in his book *On Archons*, and there they say he lived for thirty years. (2.7 = 59A1)

Even in these relatively simple statements, there is a good deal that is problematic and controversial. The first sentence of the Diogenes passage has Anaxagoras being twenty in 480 BC; the third has him being twenty in 456. The name of the archon Callias is often emended to "Calliades" to make the two statements consistent. There are, however, other ways to resolve the problem, including emending the third sentence to say that Anaxagoras stayed in Athens twenty years. So on one account, he would be in Athens approximately from 480 to 450;[5] on another, from approximately 456 to 436.[6] Another scholar has Anaxagoras living in Athens about 464–434.[7] To reconcile conflicting dates, some scholars have suggested that Anaxagoras' residence in Athens was not continuous, but divided by a hiatus.[8]

There are several considerations that influence the decisions. The fact that Socrates is never portrayed as conversing with Anaxagoras

---

5. Thus Taylor 1917; Woodbury 1981. Schofield 1980:33–35 puts Anaxagoras' *floruit* at Athens around 470–460, but more recently 1998:249 has followed Mansfeld (see next note).
6. Mansfeld 1979, 1980, followed by Curd 2007:131 and Palmer 2009:243.
7. Sider 2005:1–11.
8. Davison 1953:39–45; Cappelletti 1979; cf. Guthrie 1962–1981, 2:323.

tends to suggest that the latter left Athens earlier rather than later (that is, some time well before 430). The year 456/5 is a popular one with Apollodorus, whom Diogenes refers to as a source in his second sentence, suggesting that this is the sort of date he would use to fix Anaxagoras' age;[9] on the other hand, Apollodorus is notoriously schematic in his dating, so recognizing his dating in this might make us suspicious of its reliability as anything more than a place-holder. In any case, "Callias" appears in a sentence whose content is attributed to Demetrius of Phaleron, not Apollodorus, and clearly there is some confusion somewhere in that statement, which combines the archon year of Callias (456/5) with an age of twenty years (480, according to the first sentence of the passage).[10] Certainly Anaxagoras was the first philosopher to live in Athens, and certainly he made a deep impression on the nascent intellectual community there, where he enjoyed the patronage of Pericles and the association of leading dramatists and poets. But like much else in the fifty-year gap between the historical accounts of Herodotus and Thucydides, the particular details of his associations are lacking, and anecdotes about his life remain untrustworthy.

At the end of his stay in Athens, Anaxagoras seems to have fled prosecution:

> Around this time [the beginning of the Peloponnesian War] . . . Diopeithes passed a bill providing for the prosecution of those who did not believe in divine things or who taught theories about heavenly phenomena, which was directed at Pericles by way of raising suspicions towards Anaxagoras. . . . Fearing for him, Pericles sent Anaxagoras away from the city. (Plutarch *Pericles* 32.1, 3 = 59A17)

---

9. Mansfeld 1990:285–90 = Mansfeld 1979:60–65.
10. Mansfeld 1979:41, 57 corrects "when he was twenty years old" to "spending twenty years there," ἐτῶν εἴκοσιν <ἐκεῖ διατρίβ>ων. Marcovich 1999 even emends the last word of the sentence, "thirty" years (τριάκοντα) to "fifty" (πεντήκοντα).

There have been several attempts to identify the time of the trial or indictment, but no consensus on a specific date has been reached.[11]

As for Empedocles' life, it is even more difficult to pin down than Anaxagoras'.[12] Empedocles lived in Sicily at Acragas, a city at the periphery of the Greek world. We can say that Empedocles was an aristocrat, but he seems to have had democratic leanings and to have supported democratic reforms. One account says that he was exiled from his homeland. He is associated with no datable political events. Since his poetry stressed magic and religion as well as science, a number of legends grew up about how he stopped plagues and raised the dead. He is said to have ended his life by leaping into Mt. Etna; or, alternatively, to have faked his self-immolation.[13] All of this makes for good romance, but unreliable biography. Ancient sources are agreed in putting his activity around 470–450. We see allusions to his four-element theory in Diogenes of Apollonia and Melissus, which suggests that his physical theory was well known in the period 450–440.[14]

There are a number of doctrines that Anaxagoras and Empedocles share, suggesting some sort of influence or cross-fertilization between the two philosophers. But this does not of itself establish any priority, and further argument is needed to establish who borrowed from whom—assuming the borrowing was one-sided. So again we remain without any obvious case for priority between the two thinkers who dominated the period after Parmenides. Athenian authorities would tend to favor Anaxagoras because he lived among them. But Aristotle, an immigrant to Athens, gives an ambiguous account of relationship of Anaxagoras and Empedocles.

---

11. Mansfeld 1980. I shall not examine this problem since my solution avoids the need to worry about the issue.
12. See Wright 1981:3–17.
13. Diogenes Laertius 8.51–74 collects many of the stories.
14. Melissus B8.2; Diogenes B2. On questions of dating see Wright 1981:3–6.

It would be ideal to be able to establish the priority of one of our philosophers on the basis of evidence from history, biography, and time of publication. But such evidence is not forthcoming, and what there is, is largely ambiguous or indecisive. We have sought to explicate the history of astronomy by reference to biography without success. Perhaps a consideration of astronomical observations could add to our understanding of the historical and biographical situation.

5.2 ECLIPSES

Since the most striking accomplishment of astronomical study in the first half of the fifth century is the explanation of eclipses, let us consider what opportunities there were for observing eclipses in that period. Solar eclipses are rare phenomena. But they are in principle predictable because, as the Babylonians learned, they recur at set intervals. Furthermore, we now know in detail what the laws are that drive the heavenly bodies. Not only can modern scientists predict future eclipses, they can "retrodict" past eclipses using the same principles, that is, by running the clock backward using the latest astronomical measuring techniques and computer simulations. The problem is not quite so simple, however. The time of orbiting bodies is not constant. In particular, the earth's revolutions on its axis are gradually slowing down. Furthermore, the moon is slowing retreating from the earth in an expanding orbit. When we wish to turn the clock back thousands of years, the small changes in the time equation must be factored in. The changes are known as "delta T" (ΔT) and must be computed. Computations are calibrated by comparing results of simulations to reported ancient eclipses.[15] For our purposes the details are not important, only the recognition that in principle ancient eclipses can be determined precisely within certain limits on the basis of modern physics.

15. See Stephenson 1997.

By this procedure we can establish that two major solar eclipses were visible in the Mediterranean area during the maturity of Anaxagoras and Empedocles: one on February 17, 478 BC, and one on April 30, 463 (to use the Julian calendar).[16] The umbra of the eclipse passed over southern Greece and then Ionia in the former,[17] and over central Greece in the latter. The former eclipse was annular (the ring of the sun was visible around the disk of the moon), the latter total (the sun was completely obscured at the height of the eclipse). (Whether the eclipse is total or not is determined by the distance of the moon from the earth at the time: if it is relatively close to the earth in its elliptical orbit, a total eclipse results.) As with total solar eclipses, the maximum eclipse is visible only in narrow band along the earth's surface. The farther one is from that band, the less one sees of the eclipse until, at a certain distance, no eclipse is visible.

These eclipses would provide the only opportunities our two philosophers would have to study a solar eclipse first hand. By watching carefully, one could determine certain facts, such as that the darkened area of the sun was disk-shaped and approximately the size of the sun. The observer could use the experience to form a hypothesis about eclipses. Equally, he could test his own or another's hypothesis. By close observation and reasoning he could find problems with most of the existing hypotheses about eclipses. According to Xenophanes (as we have seen), an eclipse results from a lack of vapor to fuel the sun's illumination; but there is no obvious way to correlate a moving circular dark spot with an absence of vapor. According to Anaximenes (perhaps), an eclipse is caused by a cloud. But no cloud is so perfectly circular as the darkened area of the sun. According to Heraclitus, the eclipse results from a bowl of fire rotating to show the opaque side; but a rotating bowl would present a different configuration from that

---

16. Graham and Hintz 2007.
17. To be precise, the shadow of this eclipse was an antumbra, formed by an annular eclipse; see note 25 below.

observed.[18] As for Anaximander's theory of a blocking of an aperture, it is difficult to know how to evaluate that.

If, on the other hand, one already had the hypothesis of antiphraxis, observations would at least be consistent with the hypothesis. The disk-shaped spot could be the silhouette of the moon, an opaque, spherical body lying between the earth and the sun. The darkened area almost coincided with the size of the sun, as did the moon's own apparent diameter. Furthermore, the two eclipses occurred during the new moon, just at the time of month predicted by the hypothesis, and the only time when the moon could possibly obstruct the sun's light. Thus the two celestial events would provide an important laboratory for the new theory of eclipses. Informed observation of the events could eliminate most of the competing hypotheses and at least establish the viability of the new one. The historical eclipses could falsify most hypotheses, and confirm the antiphraxis hypothesis.

There is one set of unusual comparisons attributed to Anaxagoras by ancient sources:

> The sun exceeds the Peloponnesus in size. (Hippolytus *Refutation* 1.8.8 = 59A42)
>
> He said the sun was a fiery molten mass and greater than the Peloponnesus. (Diogenes Laertius 2.8 = 59A1)
>
> Anaxagoras [says the moon is] as large as the Peloponnesus. (Plutarch *The Face on the Moon* 932a)

So the moon is approximately the size of the Peloponnesus and the sun larger. The Peloponnesus is the large peninsula the marks the

---

18. "This does *not* 'save the appearances': the turning of a circular bowl ... makes its open side appear more and more *elliptical*, but never crescent-shaped as is the partially eclipsed sun or moon, or the moon when not full" (Kirk 1954:276, original emphasis). Mourelatos' interpretation, however, discussed above in ch. 2.2.5, would allow for a sphere of fire contained in a hemispherical bowl, and would produce the right shape.

southern extremity of Greece, with jagged coastlines, but roughly circular in shape.

Why does Anaxagoras use a peninsula of the Greek mainland as a yardstick to measure the sun and moon? One recent suggestion is that Anaxagoras used triangulation to arrive at his conclusion. According to Dirk Couprie he used a gnomon (a stick placed vertically in the ground) to measure the distance and size of the sun. First he might have measured the shadow of a gnomon at noon on the summer solstice at Delphi, which the Greeks recognized as the navel of the world. He might make a similar observation at Sparta. Assuming a flat earth one can project the base of a triangle southward to the Tropic of Cancer in Egypt, where the sun is directly overhead. By extending a line from the sun to the top of the gnomons in Delphi and Sparta and on to the ground, one gets right triangles. The decrease in the length of the shadow as one moves south allows one to estimate the distance to the tropic. The base and the base angles allow one to estimate the distance of the sun from the earth. Using Thales' estimate that the sun's diameter was 1/720th of its orbit, one can determine the approximate size of the sun. This is approximately the same size as the Peloponnesus. Couprie finds that the method gives a value of the sun's diameter as 54–78 km., depending on certain values assumed for the inputs, while the shortest east-west distance of the Peloponnesus is about 100 km.[19]

This provides an interesting deduction of Anaxagoras' results. But it has problems. The method of triangulation itself is attested for Chinese astronomers in the second century BC, but not for Greek astronomers. (Thales reportedly used triangulation to measure the height of pyramids and the distance of ships at sea in the Greek tradition,[20] but

---

19. Couprie 2006; Couprie 2011:189–200. Gamow 1964:3–5 hypothesizes a similar method for Anaxagoras, based, however, on observations allegedly made in Egypt, like those made later by Eratosthenes (he cites no sources).
20. Pliny *Natural History* 36.82 = 11A21; Plutarch *Dinner of the Seven Wise Men* 147a = A21; Proclus *On Euclid* 352.14–18 = A20.

there is no record of the method's use in early Greek astronomy.) Further, the estimate of the sun's diameter in relation to its orbit is probably the achievement of the Hellenistic astronomer Aristarchus, as we have seen (2.2.1 §14), rather than of Thales (or Anaximander, as Couprie prefers, 2006:71–2). Also on Couprie's account, it is the sun, not the moon, that is measured; but our sources rather measure the *moon*, and describe the sun's size relative to the moon. Couprie tries to impeach the sources on this point; but this is dubious methodology at best, since the sources provide the only ground we have to stand on.[21] In addition, the reference points of Delphi and Sparta seem arbitrary despite the mythological associations of Delphi, and certainly do not appear in the Anaxagoras doxography. In any case the choice of the Peloponnesus as a comparandum remains gratuitous. And the link to the Tropic of Cancer seems anachronistic, being borrowed from Eratosthenes in the third century. All in all Couprie's account is an ingenious construct, but full of unverifiable assumptions and excessively complicated.

Now in fact two scholars working independently have suggested that Anaxagoras made the Peloponnesus a measuring stick on the basis of eclipse observations.[22] M. L. West suggested that Anaxagoras used it as a reference point for his celestial geography because the Peloponnesus was shadowed by an earlier eclipse, specifically that of 557 BC. Similarly, David Sider suggested that the same reasoning led to

---

21. Couprie 2006:56–57 defends the statements as unlikely to be invented, but then he goes on to rewrite them ad libitum. He claims that the comparison of sizes between the heavenly body and the peninsula only makes sense if Anaxagoras said the sun is "a little larger" than the peninsula; yet Aetius has it "many times larger." Couprie explains the latter away as the effort of a late doxographer to bring early opinions into line with contemporary mathematical astronomy. But of course the doxographers preserve a great number of obsolete opinions, about astronomy and everything else; why is this one so embarrassing? (If Aetius' version is dubious, it is because it does not agree with other, better sources.) Couprie notes Plutarch's reference to the moon as the body measured, implying its centrality to the explanation, but then he dismisses this suggestion by questioning whether Anaxagoras (he writes "Anaximander," presumably a misprint) was in a position to measure the size of the moon (57–58). But this puts the cart before the horse, as Couprie rewrites the sources to anticipate his conclusions.

22. West 1971:233n.1; Sider 1973.

Anaxagoras' estimate of the size of the large heavenly bodies, but that the eclipse he based his calculations on was that of 463. At that time Anaxagoras was in his late thirties, probably an established thinker, and able to use the data at his disposal to make an inference about the size of the heavenly bodies. In both cases Anaxagoras must presume that the moon is relatively close to the earth and the sun relatively far away, so that the rays of the sun are roughly parallel, projecting the moon's shadow onto the earth with the same diameter as the moon itself.

This general explanation is an intriguing one. It accounts for the peculiar standard of measurement, and potentially ties in Anaxagoras' doctrines to empirical observations. But there are problems with the specific eclipses proposed. The eclipse of 557 covers the right area, but its observations would be undocumented. The Greeks kept no archives, had no journals, data-bases, newspapers, or other repositories of information. The generation that had observed the eclipse in question would be mostly deceased, and those who survived would find their memories fading. What is more, the relevant information needed to establish the extent of an eclipse is a survey of places where the eclipse was *not* visible, which could not be carried out after seventy or more years. As for the eclipse of 463, it was not visible in its totality in the Peloponnesus. It passed over central Greece, roughly Thessaly. Now it might be thought that Anaxagoras could take a compass, measure the width of the umbra to the north, and transfer this to the south, showing that the width of the shadow was approximately the diameter of the Peloponnesus. But the Greeks had no scale maps at this period, so that any displacement would inevitably introduce serious errors. The only safe way to measure something on the ground was to keep the reference point constant. Thus, both proposed eclipses seem problematic as providing the occasion for Anaxagoras' measurement.[23]

---

23. Couprie uses these points, 2006:58–59, to criticize the whole approach of using eclipses to account for Anaxagoras' measure.

If, however, we go to the eclipse of 478, we have just what we need.[24] Here the umbra (technically, the "antumbra")[25] obscured almost the whole of the Peloponnesus, as well as Athens, and indeed Clazomenae, Anaxagoras' native city in Ionia (see fig. 5.1). Anaxagoras could have seen this eclipse whether he was in Clazomenae or Athens (where, as we have seen, at least one biographical report has him after 480). Accordingly, this seems like the most promising starting point for Anaxagoras' theory of eclipses. It is the only eclipse that provides evidence for the unique view attested for Anaxagoras, which occurred during his own lifetime.

Now one problem that arises is that an annular eclipse is not nearly as obvious as a total eclipse.[26] In a total eclipse direct sunlight is completely blocked and the sky turns dark as at night, with stars and planets becoming visible. In an annular eclipse, by contrast, the sun continues to shine, though it is dimmer than usual and the sky and earth are darker than normal. How, then, could Anaxagoras view the eclipse, and in particular observe the moon's silhouette with a still-bright sun in his face? We do not have any detailed answers from early Greece, but we do get reports from later astronomers. Zeno of Citium recommends viewing an eclipse by watching its reflection in a bowl of water,[27] and earlier Plato mentioned the water method.[28] Ps.Aristotle asks about the reason why, when an eclipse is occurring, the light going through a sieve, through the leaves of a broad-leafed tree such as a plane tree, or through interlaced fingers produces a crescent on the ground.[29] The author

---

24. See Graham and Hintz 2007. Panchenko 1999:37 n. 47, cf. Panchenko 2002:333 n. 24 proposed this solution first.
25. In an annular eclipse the shadow on the earth lies beyond the focal point.
26. I was able to observe the annular eclipse of May 20, 2012, at Bryce Canyon National Park in Utah. The sun's light was noticeably weaker than before the eclipse, but the eclipse could not be observed with the naked eye, without appropriately darkened glasses.
27. Diogenes Laertius 7.146.
28. *Phaedo* 99d–e.
29. *Problems* 15.11, 912b11–14. Forster translates "through a leaf," but the Greek has the plural φύλλων, apparently indicating the dappled light filtered through the foliage of a tree. The leaf of the plane tree is quite opaque.

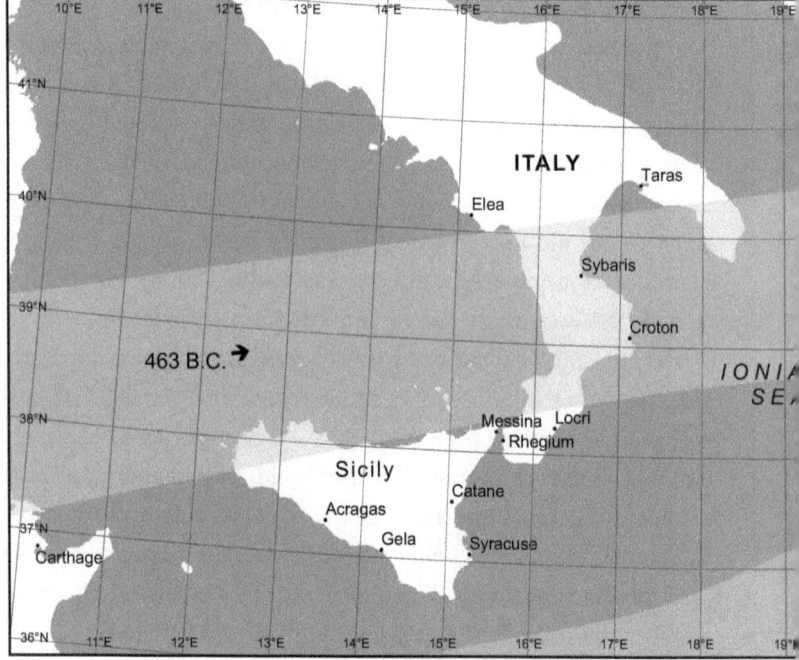

Figure 5.1. Path of Solar Eclipses: Early Fifth Century BC.

draws on optical theory to give an answer, but for our purposes, the point is to recognize that the "pin-hole camera" technique was known in ancient Greece. Finally, we have the testimony of Thucydides that an annular solar eclipse in 431 BC was duly observed and known to the public, one that was only marginally more complete than the eclipse of 478.[30] So while an annular eclipse is not an obvious event, its viewing seems to have presented no major difficulties to ancient Greeks.

Let us suppose, then, that Anaxagoras observed just this eclipse, and that it made a deep impression on him and provided the basis for his size estimates of the sun and moon. What would he have to do to

---

30. Thucydides 2.28.1, on 3 August of the Julian calendar. Gary Espenak of NASA estimates its magnitude was 0.9843 at its maximum, whereas that of 478 BC was 0.9451. See http://eclipse.gsfc.nasa.gov/SEcat5/SE-0499--0400.html, eclipses of the appropriate day of – 0477, – 0430.

DARKENED SUNS AND FALLING STARS

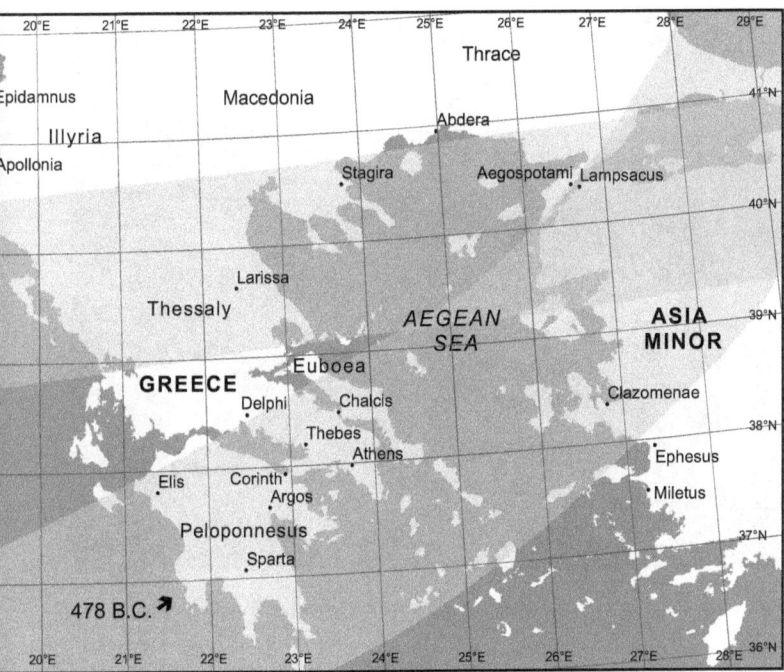

make the inference he did? As we have seen, he would have to assume that the moon was relatively close to the earth and the sun relatively far away. Thus the roughly parallel rays of the sun would project onto the earth a shadow roughly the diameter of the moon. Now we have no texts from Anaxagoras that express this assumption. But we do have a fragment from Empedocles that uses the assumption:

> ... [the moon] did away with[31] his [the sun's] rays
> to the earth from above, and it obscured the earth
> as much as was the width of the bright-eyed moon. (B42)

Here, in poetic language, Empedocles describes the moon as blocking the sun's rays to produce a shadow with the same width as the moon

---

31. Reading ἀπεσκεύασ<σ>ε; see Mesturini 1987–1988:176. Cf. Lucretius 5.753–55.

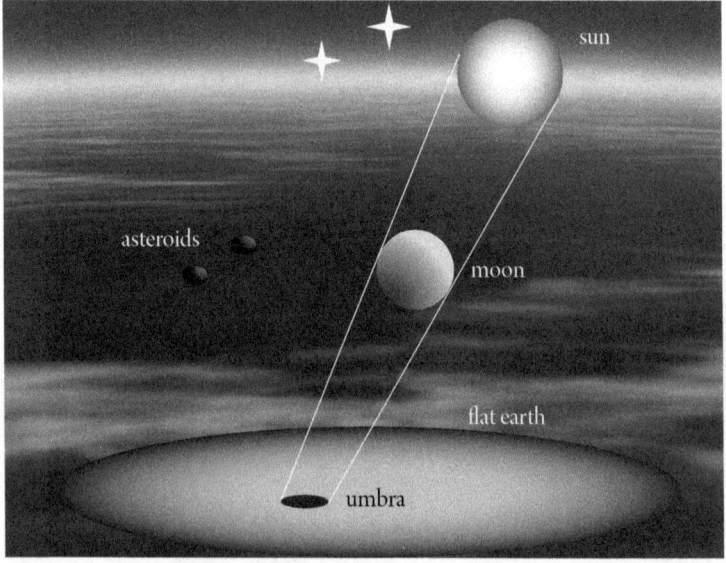

Figure 5.2. Solar Eclipse as Understood by Anaxagoras and Empedocles.

on the face of the earth. Thus we have a contemporary appreciation of the assumption needed to use the umbra of a solar eclipse as a measure of the moon.[32] (See figure 5.2.)

But it would not be enough just to make certain a priori assumptions. To measure the umbra as being equal to the diameter of the Peloponnesus is to make an empirical observation about where the maximum eclipse was visible. The observation requires reliable information—and information gathered fairly soon after the event so as to guarantee its reliability. In the first place, Anaxagoras would

---

32. This answers two of Couprie's objections (59). (1) There is no record that early Greeks concerned themselves with the width of the moon's shadow: to the contrary, Empedocles provides a counterexample. (3) The account presupposes Anaxagoras had a knowledge of the relative distances of sun and moon from earth, for which there is no evidence: but we need only very general assumptions that the moon is much closer to the earth than the sun. (Empedocles has the moon twice as far from the sun as from the earth, Aetius 2.31.1 = 31A61, following ps.Plutarch's reading.) Couprie's second objection, that the eclipse of 463 did not pass over the Peloponnesus, we have dealt with already.

Figure 5.3. Annular Eclipse, May 2012. Image of annular eclipse of May 20, 2012, taken near Nevada City, CA. Courtesy of Gpalmernc. Wikimedia. http://en.wikipedia.org/wiki/File:Annular_Eclipse,_Nevada_City,_CA_-_May_2012.jpg.

need to know where the complete eclipse was observed. But that is not enough: he would have to be able to contrast that piece of information with reports of where the complete eclipse was *not* observed. Those who saw (in refection or projection) the annular eclipse with the dark disk surrounded by a circle of fire would be duly impressed. (See fig. 5.3.) Those farther away would see less and would observe only a partial eclipse, or in some cases, perhaps, not notice the event. How then could Anaxagoras collect information? The most obvious method is that of simple interviews of travelers. He could go to a seaport and ask the sailors and merchants what they saw recently at the time of the new moon, or heard from skywatchers. This is a classic piece of Ionian *historiē* of the kind that Herodotus was practicing at about the same time: ask the locals what they know.[33] It requires no

---

33. For the roots of the concept of *historiē* see Floyd 1990; Connor 1993.

observatory, no advanced equipment, no laboratory, no national archives. One solitary inquirer goes down to the local water holes and chats up the itinerants. Interviews would suffice if the eclipse was fairly noticeable. He could have also sent letters to stargazers in other cities seeking information—a procedure that would work even if the eclipse were not especially noticeable.

But where did Anaxagoras go to get information? As we have noted, he could have seen the eclipse either from his native Clazomenae or from Athens where he came as an expatriate at some point in his career. But what he could not have done is to get information about the Peloponnesus equally well from either location. Xerxes had invaded mainland Greece in 480. His troops were there until 479, when they were routed at the Battle of Plataea. In 478, Athens was in ruins—hardly the ideal place to come on a study tour—but victorious. And she was already sending fleets to raid and harass the coasts of Persian-held Asia Minor. Ionia revolted from the Persian Empire after the Greek victory of Mycale in 479 but was threatened by Persian garrisons inland.[34] In other words, Clazomenae was in a war zone. Regular commerce and communication across the Aegean were difficult at best. It is hardly plausible to think that Anaxagoras could have conducted research on the visibility of an eclipse in the *Peloponnesus* from his native city. On the other hand, if he were in ruined Athens, all he had to do was to walk down to the port at Piraeus to interview travelers from the surrounding area and send letters to learned correspondents. Athens had come off victorious from the naval Battle of Salamis with her fleet intact and in control of the

---

34. Herodotus 9.106–107. A continuous state of war existed between the Greek states and the Persian Empire until the Peace of Callias in 449. One may assume that commerce across the Aegean was gradually restored, especially after the Greeks led by Cimon defeated two Persian fleets, one at the Eurymedon River in Asia Minor and one at Hybrus in Cyprus and then mopped up remaining Persian outposts in the Thracian Chersonese, around 466 (precise dates for this period are hard to come by). See Olmstead 1948:267–68.

Aegean Sea. Athenian trade was restored, and indeed, Athens was heavily dependent on foreign supplies to rebuild her damaged infrastructure and feed her population. Thus the Piraeus would have been an ideal spot to inquire about what people had seen in their recent peregrinations. Eclipses were still widely regarded as portents and signs among the superstitious common people and no doubt a curious researcher could get an earful by buying a thirsty sailor a drink. The time of year would have required a few weeks' delay in investigating: most shipping was suspended during winter months.[35] But warm weather would bring tidings of visitors from distant lands.

If, then, Anaxagoras' exercise in lunar measurement is based on recent data from an eclipse, he must have been in Athens, not in Clazomenae when it occurred. Now he could have conducted similar research in Clazomenae, but presumably his terrestrial reference point would not have been the Peloponnesus, but Ionia or some island or archipelago of the eastern Aegean. Thus we have reason to accept the arrival of Anaxagoras in Athens as being in the archonship of Calliades rather Callias. Why he came we do not know. He could have been a refugee from the war, an escaped hostage, or a soldier-conscript in Xerxes' force who defected or was captured. It is unlikely he came as a student at a time when the Greek world was in upheaval. Yet there were many reasons for people to be displaced in the calamities of the Persian War, and after the Persian invasion was repulsed, Athens was a safe haven and a center of political power where an intellectual might see a land of opportunity.

Whatever brought Anaxagoras to Athens, he made good use of his opportunities there. As we have noted, he needed only the simplest kind of research skills to make a survey of the area in which the

---

35. Most ships laid up from early November to early March to avoid storms and adverse sailing conditions: "during late fall and winter, sailing was reduced to the absolute minimum.... All normal activity was packed into the summer and a few weeks before and after it; at other times the sea lanes were nearly deserted, and ports went into hibernation to await the coming of spring" (Casson 1971:270–71). See Vegetius *Epitoma rei militaris* 4.39.

maximum eclipse was observed. But one thing that was essential was the hypothesis: he had already to believe that solar eclipses were the result of an occultation of the sun by the moon. The research, that is, had to be driven by a hypothesis. On *no* other theory of eclipses was there any reason to think the eclipse was visible only over a limited area. On the theory of Anaximander, the opening of the sun's ring was blocked to all observers on the earth. On the presumptive theory of Anaximenes, the sun's light was blocked by a cloud; the extent of the shadow might perhaps be measured, but since clouds are ephemeral formations, there would be nothing to learn from the exercise. On Xenophanes' theory the sun ran out of fuel in certain regions; but anyone who could see the sun should observe its disappearance. On Heraclitus' theory the convex part of the sun's bowl would block the fire of the concave part everywhere alike. Only if the solar eclipse were an occultation by the moon could the full extent of the eclipse be visible in a limited area, and that area coincide with the diameter of the moon. Only in this case would Anaxagoras have any reason to prowl the Piraeus in search of eyewitnesses from the hinterlands. In modern terms, he would have to be testing a hypothesis he had already formed.

As we have seen, Parmenides' insight of heliophotism provides the starting point for a set of implications that includes the possibility of explaining eclipses by reference to antiphraxis. If the moon gets its light from the sun, and, given that the moon is about the same apparent size as the sun (verifiable by observation), the dark moon must be in the vicinity of the sun during any new moon. If it should line up perfectly, given that it is below the sun, it would intervene between the sun and an observer on the earth, and, given that it is an opaque solid body, it would block the sun's light. Thus the heavenly mechanics derived from heliophotism presents the possibility of an occultation of the moon by the sun. If Anaxagoras already had this model in mind (as he surely did later in his career) when he observed the eclipse of 478 at the right time (the new moon) with the right sort of shape (a perfect circle) being silhouetted against a blazing sun, he would have preliminary confirmation of his hypothesis. To ascertain the extent of the shadow would be an appropriate step in celestial measurement, assuming the proximity of the moon to the earth and the remoteness of the sun (as we find in Empedocles). Heliophotism suggests antiphraxis; antiphraxis is confirmed; antiphraxis theory is used to measure the moon. A powerful theory is partially confirmed by an opportune phenomenon. An opportunistic student of the heavens uses his theory to expand his knowledge of the heavens.

We have the theory, the occasion, and appropriate data that should result from applying the theory on the occasion. We have no biographical or archival information to confirm any of this (though we shall look at some ancient corroboration shortly). The present account suggests a remarkable precociousness in a young man of perhaps twenty-two years old. Yet it requires no advanced instrumentation, no complicated mathematics, no special knowledge or privileged access to data. It requires only a familiarity with Parmenides' poem (which we know Anaxagoras had), a quick grasp of the implications of one of his astronomical insights (which Anaxagoras made at some time or other), a few auxiliary assumptions (which we know Empedocles made—or borrowed—at about the same time), an occasion on which to test them empirically, access to people from different parts of Greece, and a straightforward application of Ionian inquiry. The only unusual factor here is the occasion itself, the solar eclipse, which we can, by the use of modern computer models, verify as happening on February 17, 478. The reference to the Peloponnesus suggests the researcher's location (among limited possibilities). The reference point could in principle have been filled by some other place had the researcher had access to different data. Thus we can offer a consistent and historically plausible account of the when and the where of the discovery of antiphraxis, when a speculative hypothesis received powerful empirical support and was tested for further applications: February 17, 478, Athens.

As for the who, it is fairly obvious: Anaxagoras. At this time he was a young adult. Empedocles, who was younger, presumably by five to ten years (based on Theophrastus' testimony, cited above), would have been too young to be part of the adult intellectual world at say twelve to seventeen years of age. (He would still have been under the governance of a *paidagōgos* or tutors.) Furthermore, he could not have seen the eclipse first-hand, since the umbra did not pass over Acragas in Sicily. Nor would he have had the freedom to conduct research at the adjacent seaport, nor, presumably, would he have had enough visitors from faraway Greece to plot the path of the umbra, even had he been aware of the event. In fact, he would not have been able to observe a full solar eclipse in his lifetime from Sicily (since the eclipse of 463 also passed his home by). Thus it seems unlikely on the basis of age, location, and opportunity that Empedocles was able to test the hypothesis of antiphraxis properly. That he grasped its importance is evident from his fragments. But that he originated or verified the theory is unlikely. Anaxagoras emerges as the great astronomer of the early fifth century, and as one who made his most important advances at the very beginning of his career.

One further piece of empirical evidence would have emerged from the eclipse: since the eclipse was annular, it was evident at a glance that the sun was larger than the moon.[36] This point is made in the doxographic reports: the moon is the size of the Peloponnesus, the sun is larger.[37] Anaxagoras' opinions follow from the model

---

36. Sider points out that Anaxagoras had a theory of perspective, which he thinks accounts for the judgment that the sun is larger than the moon: Sider 1973; 2005:10, 18–19, 88; with Vitruvius 7, pref. 11 = 59A39. But this leads Sider to say that the umbra of an eclipse would be *smaller* than the moon, which is not what the Anaxagoras seems to have thought. In any case, if the eclipse he studied were annular, he could see that the moon was smaller than the sun by *inspection*. An understanding of geometric perspective could add to observation in a secondary way.
37. This provides an immediate motivation for the point that troubles Couprie: why the moon rather than the sun provides the point of departure for the measurements.

and the occasion. If Parmenides had gotten some facts of astronomy right for the first time, Anaxagoras was making empirical tests of astronomical theories and drawing some correct conclusions from them. That he was able to do this presupposes that he had already tested heliophotism against the phases of the moon. Now, however, he took advantage of a once-in-a-lifetime opportunity in seeing one of the rarest and most spectacular of heavenly phenomena, the solar eclipse,[38] to test the implications of heliophotism in a new way: to see if it could explain the phenomenon itself. One of the implications of the theory was that the full eclipse should occur only at the new moon; another that the sun's light should be blotted out in a circular pattern; these implications could be verified immediately. Another was that the full eclipse should be visible in only a limited area. This, it appears, Anaxagoras discovered by an additional effort. Here is the first case of something like a case of collecting data to confirm an empirical hypothesis of astronomy by some method other than close observation of the phenomenon itself. This looks like real empirical science.

5.3 THE METEOR

Anaxagoras is famous for one other spectacular success in astronomy:

> They say he predicted the fall of the stone which occurred near Aegospotami, which he said would fall from the sun. (Diogenes Laertius 2.10 = 59A1)

There is no doubt the event took place. It provided an important headline in chronicles of the times[39] and the meteorite itself became a tourist

---

38. At any given place on the earth's surface a solar eclipse occurs only about once every three hundred years, on average.
39. Marmor Parium 57 = 59A11.

attraction for at least five centuries after.[40] The report is puzzling, because we know that meteors are not predictable (except for certain annual meteor showers). The report has the appearance of a legend like that of Thales' prediction of a solar eclipse. Yet, like Thales' statement, this event is well documented, and must have some basis in fact.

Writing in the first century AD, Pliny the Elder gives a more detailed account:

> The Greeks report that Anaxagoras of Clazomenae, in the second year of the 78th Olympiad [467/6] predicted, by his knowledge of astronomical writings, within what days a stone would fall from the sun, which happened in the daytime in a region of Thrace on the Aegos river. The stone is still exhibited, being about the size of a wagon load, of a burnt color, which fell while a comet was seen burning in the night. (*Natural History* 2.149 = 59A11)

Although this account is late (first century AD), it is confirmed by an excellent early source:

> When the stone fell from the air at Aegospotami, having been supported by the wind, it fell down during the day. It so happened that at the same time a comet was visible in the west. (Aristotle *Meteorology* 344b31–34)

Aristotle connects the meteor with winds, in accordance with his own meteorological account of stones falling from the sky. But he confirms that a meteor fell during the daylight hours near Aegospotami (Goat Rivers) in northern Greece, at the same time a comet was visible, specifically in the western sky (see map in Fig. 5.1, p. 151 above; Aegospotami is just across the Hellespont from Lampsacus, where Anaxagoras retired).

---

40. Pliny *Natural History* 2.149 = A11; Plutarch *Lysander* 12.1–2 = A12.

The remaining meteorite was the size of a wagon. Since most of a meteoroid disintegrates on contact with the earth (after much of it has been burned up by friction with the atmosphere, the original stone itself must have been huge, and the impact must have left a sizable crater.[41] Pliny repeats the apparently early report that Anaxagoras predicted the eclipse; he adds the details that he predicted the fall within a limit of several days, and that he based his prediction on "astronomical writings" *caelestes litterae*. But such Greek writings as there were would have treated heavenly bodies as light bodies or meteorological formations, not as stony or earthy, while Babylonian writings did not deal with the composition of heavenly bodies. Hence it appears that on this point Pliny is guessing or following a poor source.

Writing a generation or more after Pliny, Plutarch gives his own assessment:

Some say the fall of the stone was an omen of this disaster [the Athenian naval defeat at Aegospotami, 405 BC], for an immense stone had fallen on Aegospotami, according to common belief. (It is shown to this day by the inhabitants of the peninsula, who stand in awe of it.) It is said that Anaxagoras predicted that one of the bodies entangled in the heaven might, if there were some slip or agitation, break off and fall or be cast down; and indeed none of the stars is in its natural place; being stony and heavy they shine by resistance of the revolving aether, and being constrained by the angular momentum of the revolution they are dragged by force, which is why when they were at first captured

---

41. Meteor Crater in Arizona, 570 (170 meters) feet deep and 4100 feet (1.25 km.) wide, is thought to have been caused by an asteroid 130 feet (40 meters) wide. The largest remaining fragment, the Holsinger Meteorite, composed of iron and nickel, 639 kg. in weight, could count as a wagon load in weight, though smaller in size than a wagon. See http://en.wikipedia.org/wiki/Meteor_Crater.

they did not fall down here, as the cold and heavy things were being separated off from the totality. (*Lysander* 12.1–2 = 59A12)

Plutarch is familiar with Anaxagoras' physical theory, and uses it to explain the prediction. What Anaxagoras predicted, according to Plutarch, was not the specific fall of a meteor at a given time, but the *possibility* that a rocky body could fall from the sky—given the nature of heavenly bodies he hypothesized. In terms of Aristotelian science, the heavenly bodies are not in their natural place, but are held aloft by the force of the cosmic vortex. If there is some disturbance, one of the stony bodies could fall to earth. As we have seen, Anaxagoras posits the existence of asteroids below the moon, which would presumably be held more weakly than those bodies farther out where the vortex motion is stronger; hence they would be unstable and liable to fall out of orbit. All of this is, according to Plutarch, a perfectly general point. When a stone happened to fall at Aegospotami, one can say that the generic prediction was fulfilled.

What Plutarch offers us is a kind of philosophical reconstruction. But it is based on whatever information he had on Anaxagoras and on his own wide knowledge of philosophical and scientific literature.[42] Plutarch presents an attractive reading of the evidence, one that does not require us to attribute to Anaxagoras supernatural powers of prediction or access to fantastic astronomical lore. All in all, it seems quite level-headed and plausible.[43] Apparently this minimalist account fit with the doxographical evidence Plutarch had at his disposal and allowed him to view the prediction as a theoretical entailment rather than a customized forecast.

---

42. Plutarch's dialogue *The Face on the Moon*, for instance, shows that he was conversant with both ancient philosophical theories and recent scientific models of astronomy.
43. Stokes 1965:226–27 accuses Plutarch of "carelessness" as well as of being unclear in his account. Citroni Marchetti 2007:130–31 also finds Plutarch unclear. It seems to me, on the contrary, that Plutarch's account is both careful and clear.

Plutarch seems to have the right approach to the problem. What was unique about Anaxagoras' theory was its depiction of heavenly bodies as solid, heavy bodies, rather than light, thin textures that remained aloft by their natural lightness or by some minimal wind activity, as does a kite. When a large stone fell out of the sky, his theory was vindicated against all competing theories. Only his theory posited heavy bodies; only his theory predicted that stones *might* fall from the heavens. His theory, consequently, was confirmed and other theories falsified by the meteor of Aegospotami.

All of this is fairly straightforward. Yet we should pause to consider how remarkable it is that the meteorite was associated with the name of Anaxagoras. A generation or two earlier, and indeed even in 467, one would expect that the common people would view a meteor as a divine portent of some sort. According to Hesiod, the Hundred-Handers threw great boulders at the Titans in a primeval war.[44] A large stone from heaven should be caused by some great Olympic deity. Yet we hear of no mythological connections.[45] On the other hand, the event was connected to a philosophical theory. No report tells us that Anaxagoras explained the event ex post facto. Rather he predicted it. His theory evidently was in the public domain, and well-known at that, if it immediately became associated with a heavenly prodigy. The report is like an ancient headline: ANAXAGORAS' THEORY VINDICATED.

The report implies that Anaxagoras' theory was known well beyond the boundaries of Clazomenae or Athens by 467. The most likely reason for this is that his book was available and in circulation, at a time when the philosopher was around thirty-three years old. Now the story of the solar eclipse provides a terminus post quem for at least part of his theory: after 478 Anaxagoras would have the data

---

44. *Theogony* 671–86, 713–21.
45. Cf. Homer *Iliad* 4.75–77, quoted in the epigraph to this chapter.

to "prove" that the moon was the size of the Peloponnesus and the sun was larger. More importantly, he would have the evidence of the eclipse to prove that the moon was indeed an opaque spherical body capable of blotting out the sun. Thus there was at least one massy body in the heavens, and by an inductive generalization, it seemed plausible that all heavenly bodies were heavy earthy or stony objects. He seems to have needed asteroids to complete his account of lunar eclipses, so he had asteroids in his theory. None of this theory is dependent on the meteor of Aegospotami in any way. The theory seems to have been worked out independently of that event. But once the meteor fell, the event became the unexpected but indubitable empirical confirmation of his bold theoretical construction: here was irrefutable evidence that there are heavy bodies in the sky. All earlier theories were refuted. Only Anaxagoras' remained as a viable account of heavenly phenomena. It accounted for the phases of the moon in a striking way that no earlier theory had come close to; it explained eclipses and made theoretical predictions about the time of month they had to occur, one of which had been verified; it posited the existence of heavy bodies in the sky (contrary to all earlier theories), and this implication was now verified in a spectacular way. The Lithic Model was vindicated for all to see.

For the first time, indeed, there was indisputable evidence on the ground of massy bodies existing in the sky. The sensation this caused must be comparable to the display of moon rocks in museums in recent times. The heavens were brought down to earth. More important, the meteor's fall ushered in a new era for astronomy. Those who might have said that philosophical theories were just airy speculations could run their fingers over a piece of the heavens. Some theories were more than mere speculations. They made risky predictions that on rare occasions could be verified.

The meteorite at Aegospotami provides an important landmark in Anaxagoras' career. Does it tell us anything about his location? In principle he could have developed his general theory in either Clazomenae or Athens (although we have seen reasons why his measurement of the moon makes sense only in Athens). Yet, given the ongoing conflict between the Greeks and the Persian Empire, it is easier to think that his reputation and his book would circulate more easily from Athens. Ionia was cut off by the rising power of the Delian League, directed by Athens, while Athens sat at the center of a vast trading and military network that reached into the Thracian Chersonese where Aegospotami was. Athens controlled the trade routes of the Aegean and was in communication with the cities of the Hellespont. It would not have been impossible for Anaxagoras' book and his reputation to spread from Clazomenae, but it would have been more difficult.

5.4 THE COMET

There is one further point about the meteor that calls for comment. Pliny mentioned that the meteor fell while a comet was burning in the sky. Anaxagoras had a theory of comets:

> Comets are a conjunction of planets releasing flames. (Diogenes Laertius 2.9 = 59A1)
>
> Now Anaxagoras and Democritus say [comets] are a conjunction of wandering stars [i.e., planets] when as they come near they seem to touch each other. (Aristotle *Meteorology* 342b27–9 = 59A81)
>
> Anaxagoras and Democritus say [comets] are a conjunction of two or even more stars joining their radiation. (Aetius 3.2.2 = 59A81)

It is not clear from Diogenes and Aetius whether the planets that cause the appearance of comets touch or not; Aristotle more explicitly says they do not really touch, but only seem to. If this is the case, then there is nothing more than a joining of light from two bodies not in contact, and a comet is a mere optical illusion of sorts.[46] On this theory the presence of a comet would have nothing to do with a meteor, and the two events juxtaposed in Pliny's account would be mere coincidental occurrences; consequently the observation of a comet could not support the prediction of a meteor.

On the other hand, Anaxagoras' account of a conjunction of planets (whatever his precise words) could have another significance. If any body that does not travel in company with the fixed stars counts as a planet (a "wanderer," as often the sun and moon so count) then asteroids too might be reckoned as planets. And if two asteroids came into close proximity, they might cause friction and suddenly become visible despite the fact that they are normally dark bodies.[47] And this kind of close encounter could cause one of the involved bodies to fall out of orbit; or it might interfere with the equilibrium of some nearby body, in the manner suggested by Plutarch's account of astronomical disturbances. In this case the appearance of a comet might "predict" a meteor in a more specific way than the general theory of the heavens did.

Pliny's account raises one more intriguing possibility. Sir Isaac Newton's friend Edmond Halley calculated the orbit of a comet that appeared in his lifetime and when it returned on schedule—long after Halley's death—it became a symbol of the power of Newtonian science. That same comet has been cycling about the sun with a period of about 76 years since time immemorial. Given the timing of the events described in Pliny, the comet that appeared at about the time of the meteor could be Halley's comet, the first recorded

---

46. This is how Alexander of Aphrodisias understands the theory: Alexander *Meteorology* 26.11–16 = 68A92. The conjunction of two of the five planets (he names them all) produces a *phantasia* of a single body. The naming of the five planets seems anachronistic for Anaxagoras and probably Democritus as well (Seneca *Natural Questions* 7.3.2 = 68A92).

47. This is the view of Burkert 1972:311 and n. 66.

sighting.[48] According to two independent calculations based on computer models, Halley's comet should have reached its perihelion, or closest approach to the sun, around July 18 of 466.[49] At this time the comet would have risen above the plane of the solar system about the time it crossed the earth's orbit, and it would have appeared between the sun and the earth, making a close passage to the earth (0.46 astronomical units [AU], where an AU is the mean distance from the sun to the earth) while its tail passed over the earth (see figs. 5.4, 5.5).[50] Indeed, Plutarch represents just such an encounter, in the continuation of his report:

> In his treatise *On Piety* Daïmachus supports Anaxagoras,[51] reporting that before the stone fell, for seventy-five days a huge fiery body was visible in the sky, like an inflamed cloud, not still but moving with complex and branching motions, so that fiery fragments from its shaking and errant course flew in every direction, flashing like shooting stars. But when it had fallen to earth there

---

48. This possibility was suggested by Schove 1948:181 and West 1960. Kamienski 1956:127–28 reports that M. A. Viliew in a little-known Russian publication in 1917 calculated that comet Halley made a return in −465.73 (i.e., late in 466 BC); Kamienski makes his own calculations in the same article and in Kamienski 1957.
49. Yeomans and Kiang 1981. Landgraf 1984:35 identified the comet in question with comet Halley, but withdrew his identification in 1986:258–59, when he revised his calculations. Ironically, his new calculations (using different parameters and methods of calculation) put the perihelion of the comet within 8 hours of Yeomans and Kiang's estimate—at a distance of 226 years from the earliest previous confirmed sighting! Landgraf thought the identification failed because the estimated perihelion now fell outside the limits of the year given by Pliny and also by Chinese records. But Eusebius' *Chronicle* puts the comet in 466/5 (1551st year of Abraham = Ol. 78, 3, p. 110 Helm). For texts on the comet, see Ramsey 2006:52–56 (as Ramsey observes, "none of [the sources is] noted for chronological accuracy," 52). There is a report of a comet sighting by Chinese observers for 467 BC (more precisely December 468–December 467), but the Chinese records for the period in question are good only to about ± 1 year (on this see Stephenson 1990:241–42), and so do not exclude summer 466. For his part, Stephenson, though he thinks Yeomans and Kiang provide the best model for the comet (250, 232–3 and fig. 13.1), notes that other studies fail to agree with it; yet he refers only to Landgraf's 1984 study, ignoring the corrected report of 1986 that agrees closely.
50. Graham and Hintz 2010.
51. Daïmachus seems to have flourished in the early third century BC and when he served as an ambassador for Antiochus I; see *Oxford Classical Dictionary*, 3rd edn, s.v. Daimachus.

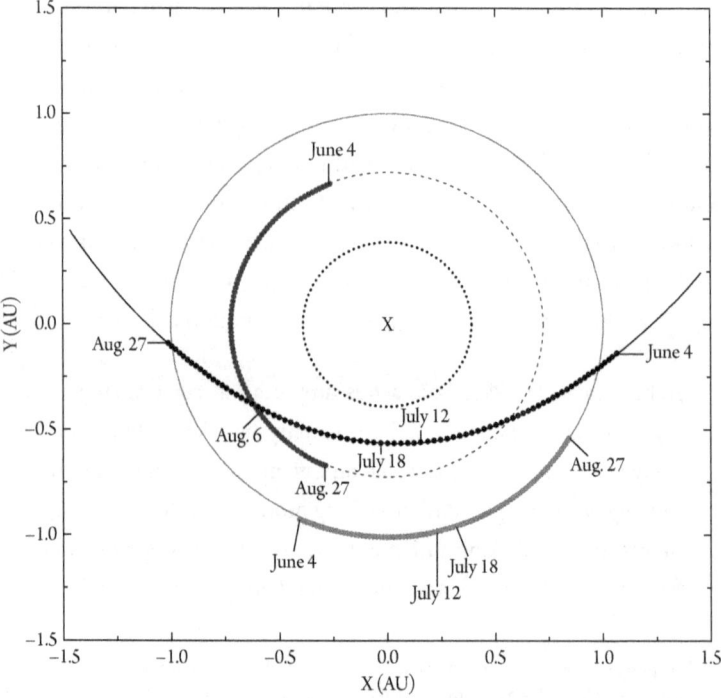

Figure 5.4. Halley's Comet. Path as calculated for 466 BC. Light gray = path of earth; dark gray = path of Venus; black = path of Comet Halley. Calculated by Eric Hintz (see Graham and Hintz 2010).

and the local inhabitants got over their fright and gathered around it, they saw no activity of fire, not even a trace, but a rock lying there, large indeed, but representing no appreciable fraction of that fiery mass above. (Plutarch *Lysander* 12.4–5 = 59A12)[52]

52. See also Seneca *Natural Questions* 7.5.3 = A83: "Charmander also, in that book he wrote about comets, says a great and ominous light the size of a massive beam was seen by Anaxagoras and shone for many days." Bicknell 1968b interprets the phenomenon reported to be an aurora, which could only be visible at the latitude of Greece if caused by sunspot activity. He further supposes that Anaxagoras had seen a sunspot and accordingly thought that pieces of the sun's surface had broken off. This interpretation fails to account for the 75-day appearance of the phenomenon and its confirmation as a comet by Aristotle and his source.

Figure 5.5. Comet with Elongated Tail. Comet McNaught, by Akira Fujii, Australia, 2007; http://www.pbs.org/seeinginthedark/astrophoto-gallery/cometmcnaught.html. In the reconstructed path of Comet Halley in 466 BC, the comet's tail would extend above the north pole and beyond the earth.

Seventy-five days is a long time for a comet to be visible, but not unprecedented. Calculations show that Halley's comet would have been visible for up to eighty-two days during the reconstructed passage. The shooting stars Plutarch reports from Daïmachus' account are not a regular part of cometary appearances, but could be explained as the earth's passing through the debris field of the comet, which is likely to have happened according to the reconstruction.[53]

53. Ramsey 2006:53, 2007 points out that the evidence we have cannot definitely prove that Halley's Comet was Anaxagoras' comet, which is quite right. Yet his main reason for remaining noncommittal is the claim that reconstructions have failed to reach a consensus. As we have seen, however, two of the most careful studies, using different methodologies, agree very closely; see note 49 above. While the identification of Anaxagoras' comet with comet Halley cannot be demonstratively proved, the confluence of evidence makes the identification at least plausible. For more on the comet, see Graham 2013.

Plutarch regards the meteor as a possible fragment of the comet, and hence as causally related to it. This is consistent with seeing the comet resulting from a collision of asteroids in Anaxagoras' theory. Yet the theoretical reports we get, strongly supported by Aristotle, suggest that Anaxagoras rather saw comets as the result of a conjunction of planetary light. Given that the appearance of the comet was so impressive (and in fact accompanied by a couple of planetary conjunctions, which it would have outshone),[54] it seems likely that when Anaxagoras proposed his theory of comets he had never seen a major appearance. This tends to support the view that his theory was proposed and published prior to 466.

5.5 THE NILE FLOODS

One more theory of Anaxagoras' is important for questions of dating. This is his theory of the Nile floods. Although it is not an astronomical theory, it is unique to Anaxagoras, and it will help to locate the period of his theoretical activity.

The Nile river flooded in the summer, usually after the summer solstice.[55] This seemed unusual to Greek theorists because all the rivers in their lands flooded in the spring, obviously as a result of melting snows in the mountains that fed the rivers. Egypt, however, is a hot, dry land, and the Nile that supplies its water flows from tropical lands to the south. How could it flood, especially in the middle of summer? Thus, while flooding rivers in Greece found a ready explanation in natural philosophy, the Nile floods did not.[56]

Herodotus takes up this question, and reviews three Greek theories. The first theory of the Nile comes from Thales, who held that it

---

54. See Graham and Hintz 2010.
55. The flow of the river is now regulated by the Aswan High Dam.
56. Herodotus 2.19.

was to be explained on the basis of the etesian (or "annual") winds.[57] The etesian winds start to blow in the Mediterranean about the time of the summer solstice.[58] They blow strongly from the north for much of the later summer. As we now know, they are generated by the heating action of the Sahara Desert. As hot desert air rises it creates low pressure which sucks in masses of cool air from Europe. Thales argued that the powerful winds blowing opposite to the current of the Nile caused its waters to back up and hence to flood. Thus he drew on special weather conditions that coincided with the floods to explain them.

The second theory was that of Hecataeus.[59] He believed in Ocean, the waters that flowed around the rim of the flat, circular earth, and was a part of mythical geography in Homer and Hesiod, and retained in Anaximander's map. Hecataeus held that the source of the Nile was Ocean, and the river's summer floods results from an increased flow from the source.

The third theory was that of Anaxagoras.[60] He held that the floods were caused by melting snows, just as in Greece, but in this case the snows were in high mountains to the south. In fact, the Nile is fed by melting snows from the south. But that is not what causes the floods: the floods result from monsoon rains that fall in the Ethiopian highlands in the spring, generated by weather patterns in the Indian Ocean.[61]

What is significant from the standpoint of dating is that Anaxagoras' theory was known to all of the great Athenian dramatists:

---

57. Herodotus 2.20. Herodotus does not name the authors of the three opinions, but we can discover them from other testimonies. For Thales, Seneca *Natural Questions* 4a.2.22, Aetius 4.1.1 = 11A16.
58. Aristotle *Meteorology* 361b35–362a11; he identifies the beginning of the winds with the rise of the Dog-star (Sirius), i.e., in late July. I can testify from experience that sailing in a boat and landing in a small airplane in the etesian winds can be dangerous undertakings.
59. Herodotus 2.21; Hecataeus fr. F302 Jacoby with Lloyd 1976:100.
60. Herodotus 2.22; Hippolytus *Refutation* 1.8.5 = A42; Aetius 1.4.3, Seneca *Questions on Nature* 4a.2.17 = A91.
61. See Bonneau 1964:15.

Aeschylus, Sophocles, and Euripides. It is Aeschylus' knowledge that interests us. A fragment from one lost play says:

> I have learned to praise the race
> of the Ethiopian land, where seven-streamed Nile
> rolls with rain of winds on earth,
> where sun beaming out fiery flame
> melts the mountain snow, and all thriving
> Egypt, flooded with the holy stream
> raises the life-giving ear of Demeter. (fr. 300 Radt)

Aeschylus alludes to this theory at least three times in lost plays.[62] Ancient sources all agree he got this theory from Anaxagoras, the only philosopher to propose such an account. Now Aeschylus died in the year Callias was archon, precisely when, on one account, Anaxagoras came to Athens. We know about what Aeschylus was doing in his last years: presenting the *Oresteia* and traveling to Sicily. Whenever he wrote the lost plays that allude to snow-melt theory of the floods, it was before 460. Now it is possible that Anaxagoras was well enough known that his theories captured the fancy of intellectuals and playwrights before he made his way to Athens. But on the other hand, given that Aeschylus was not a philosopher, it seems at least as likely that he got the theory from the horse's mouth. Whereas Euripides was interested in philosophical and sophistical theories of all kinds, Aeschylus was not.[63] The veteran of Salamis seems to have drawn on the cultural life of Athens, but not on the Panhellenic philosophical movement. As for his possible acquaintance with Anaxagoras, one can at least suggest an obvious

---

62. Besides the cited fragment, also *Suppliants* 497, 561. Also cited in Sophocles fr. 797, Euripides *Helen* 1–3 and fr. 228.
63. He seems to have had some interest in observational astronomy, but not specifically in philosophical accounts of it: Pfundstein 2003.

connection: Pericles produced an early play of Aeschylus, and Pericles was Anaxagoras' patron as well as Aeschylus'.[64] The playwright's familiarity with and favorable attitude to the theory tends to support Anaxagoras' early arrival at Athens.

We know about the three early theories of the floods in part because the question exercised Herodotus. Indeed, it appears that the young historian had the philosophical theories in mind when he interviewed Egyptian priests about the causes of the floods.[65] He was disappointed in their answers—which he does not bother to repeat—perhaps because they were couched in terms of the Nile god or other non-scientific considerations. He does not mention any other theories of the Nile, though Diogenes, Democritus and others addressed the same problem later. It appears that Herodotus' account reflects the state of research in the early fifth century when he was a young tourist in Egypt. We do not know exactly when he was there, but since Egypt and Libya revolted against the Persian Empire in 460, making travel there dangerous from then until the suppression of the revolt in 449,[66] we may suppose that his visit came earlier. Thus Herodotus too seems to provide evidence that Anaxagoras' theory was known by the 460s. Since Herodotus was from Halicarnassus in Ionia, he could have gotten it from a neighboring Ionian, which is consistent with Anaxagoras' being in his home city. But in any case Anaxagoras' theories seem to have been known from one side of the Aegean to another by the end of the 460s.

All of this makes it unlikely that Anaxagoras' writings were published late in his career as Aristotle's remark (taken literally) suggests. He was the darling of the Athenian intelligentsia and the philosopher par excellence of the 470s and 460s. The fall of meteors called his name to mind, and tragedians in Athens and educated travelers in

---

64. *The Persians*: IG II² 2318 with Podlecki 1998:11, 17–34.
65. Herodotus 2.19. See Graham 2003b.
66. See Olmstead 1948:303–10.

Egypt had his theories on their lips. In the absence of newspapers, journals, and other modern forms of communication, it seems likely that his views were published in a book sometime before 460, a result that agrees with our conclusions about the date of his astronomical theory.

CONCLUSION: THEORY AND EVIDENCE

Beginning with Parmenides' insight of heliophotism, astronomical theory began to suggest evidence for its own verification. However Parmenides came to his insight, he and anyone who was aware of the insight could test it against the monthly phases of the moon. It turns out that the phase of the moon correlates with its angular distance from the sun, with the luminous part of the moon's sphere always turned towards the sun at its closest appearance (whether to the west or the east of the moon). Empirical observations thus support a causal connection: the moon is indeed illumined by the sun's rays.

As we saw in the last chapter, heliophotism has a series of implications: (1) the moon is opaque; (2) it lies below (nearer to earth than) the sun; (3) it is spherical or spheroid; (4) the sun and moon are permanent bodies; (5) the heavenly bodies are massy; (6) the paths of the sun and the moon travel under the earth; and (7) eclipses *might* be caused by antiphraxis, blocking of the sun's light. Anaxagoras and Empedocles recognized and accepted all of these implications, and went further to establish that eclipses were in fact the result of antiphraxis. The most striking evidence for the last implication came in the solar eclipse of 478, which only Anaxagoras was in a position to study. He seems to have conducted a piece of *historiē*, collecting data on sightings of the event, to measure the antumbra of the eclipse and from this to infer the size of the moon and the relative size of the sun. Thus the eclipse of 478 provides the occasion when a hypothesis about eclipses received empirical confirmation and became a substantiated theory.

Apparently because of his need to account for lunar eclipses that happened when the moon was rising or setting, and because he (like almost all of his predecessors) thought the earth was flat, he posited the existence of asteroids, small bodies that were earthy or rocky like the sun and moon, and like them were held in orbit by a powerful vortex motion, in accordance with the Lithic Model.

Anaxagoras' theory seems to have been in place after 478 and before 466. At the latter date a meteor fell from the sky and left a large meteorite near Aegospotami. Because the theory of asteroids was unique to Anaxagoras' theory and was known antecedently, the meteor was said to have been predicted by the philosopher.

Thus we can say that astronomical theory as originated by Parmenides and developed by Anaxagoras found three separate kind of empirical confirmation: (i) heliophotism can be verified by close observation of the moon's monthly cycle; (ii) the related but more theoretically rich account of eclipses was verified by the eclipse of 478 BC, which satisfied the prediction of occurring at the new moon and which was found to have a shadow of finite diameter, as predicted; and (iii) the further corollary of asteroids was verified by the stone of Aegospotami. Both the solar eclipse and the meteor were events that were fortuitous in that there is no guarantee that either a solar eclipse or an observable meteor impact will happen in a lifetime of study. Yet both occurred, and both were used to verify the complex of theoretical implications that arose from Parmenides' insight. Heliophotism, antiphraxis, and the solid composition of heavenly bodies had been proved empirically. Anaxagoras did not invent heliophotism, but he exploited it early in his career, and through fortuitous events was able to have his theory vindicated in spectacular occurrences. All of this happened in the early years of the fifth century and in the Aegean, far removed from Sicily, making it highly improbable that Empedocles anticipated Anaxagoras in the publication of a theory of the heavens embodying heliophotism and antiphraxis. In all likelihood Anaxagoras was the first thinker to apply Parmenides' insight to produce a comprehensive theory of the heavens with testable consequences.

In ch. 3 we reviewed five questions:

1. Who discovered the theories in question first?
2. What led him to this discovery?
3. Did Anaxagoras and Empedocles have good evidence for their astronomical theories?
4. Did the community of philosophers accept the theories?
5. Did Greek theorists of the fifth century develop the correct theory of eclipses on their own, or did they borrow it from another source?

In answer to question (1) we have already seen (ch. 3) that Parmenides proposed heliophotism. In this chapter we have seen that Anaxagoras is in all likelihood responsible for antiphraxis. As to (2), we have seen (ch. 4) that Anaxagoras probably elaborated his theory as a series of applications of heliophotism to astronomical problems. In the present chapter we have seen that the question of evidence (3) can be answered in the affirmative. Anaxagoras had preliminary empirical evidence in the phases of the moon, and he gained further empirical confirmation in the eclipse of 478 and the meteor of 466. As to the question of provenance (5), we have seen no need to invoke any major influences earlier than Parmenides or farther afield that Elea and Clazomenae. We have not yet spoken of the reception of the new theory (4); to that question we now turn.

Chapter 6

Lunar Revolutions

The Triumph of the New Astronomy

Hypotheses
1. That the moon receives its light from the sun. . . .
3. That, when the moon appears to us halved, the great circle which divides the dark and bright portions of the moon is in the direction of our eye. . . .
5. That the breadth of the (earth's) shadow is (that) of two moons.
—(Aristarchus *On the Size and Distances of the Sun and Moon*, trans. Heath)

This [namely, parallax] is the reason why in the case of solar eclipses, which are caused by the moon passing below and blocking [the sun] . . . [in contrast to lunar eclipses] the same [solar] eclipse does not appear identical, either in size or in duration, in all places.
—(Ptolemy *Almagest* 4.1, trans. Toomer)

We have found reasons to believe that the major reconceptions of astronomy in the early fifth century BC resulted from the work of one man, Anaxagoras, building upon the insight of another, Parmenides. We have seen how theoretical considerations could lead Anaxagoras to many of his opinions. We have also seen that some fortuitous astronomical events helped provide timely empirical evidence for what otherwise might have remained an idiosyncratic and mostly speculative theory. It has been important to examine how the theories of heliophotism and antiphraxis arose. Yet it is equally important for our purposes to see what happened to them after they were set forth. The received view of Presocratic theories is that they remained merely speculative hypotheses, precociously brilliant but unproved and unprovable with the small means at the thinkers' disposal and their imperfect conception of the need for empirical confirmation. Only in the fourth century BC at the earliest, it is claimed, were students of the heavens in a position to build mathematical models and test hypotheses about the heavens. Until that time, natural philosophers were guessing, without the possibility of producing adequate evidence in favor of their theories or persuading their peers. If there was no process for persuading others by appropriately scientific arguments, there was no real science. If, conversely, there was real science, there must have been some way to persuade a community of researchers that some theories were right.

If Anaxagoras arrived at the correct account of the moon's light and of eclipses, could he persuade anyone else of his success? Or were his theories doomed to become just another exhibit in a menagerie of exotic theories among which one could not by rational and empirical means choose until further tests had been developed?

6.1 A COMMUNITY EFFORT

Science is a social enterprise that requires a community of researchers, as C. S. Pierce understood.[1] A lone individual arriving at an important idea is not making a scientific advance until the idea has been put to the test and proved to the satisfaction of a community of individuals

---

1. "The opinion which is fated to be ultimately agreed on by all who investigate, is what we mean by the truth, and the object represented in this opinion is the real." "How to Make Our Ideas Clear."

who share some common understanding of what they are looking for and how to know when they have found it. This is not to say a new idea cannot constitute a scientific hypothesis or truly describe the world. But in the absence of some procedure for testing, it is still just an interesting idea. As Xenophanes says, "even if [one] should completely succeed in describing things as they come to pass,/nonetheless he himself does not know: opinion is wrought over all" (B34.3–4). Thus Anaxagoras could in principle hit on important truths about astronomy and yet fail to make a scientific advance if he could not convince others of his advance, and convince them using appropriate arguments and evidence. What science requires is some public space for testing ideas.

Let us take a very simple example of a question of discovery. Who discovered America? In recent times critics have complained about the formulation of this question, as implying that the American continents and its peoples needed to be discovered. But the point the question asks is, Who from the Old World or the Eastern Hemisphere revealed the existence of a New World or the lands and peoples of the Western Hemisphere? Often the debate has taken place simply on the basis of who arrived here first from the Old World. Probably Leif Ericson and others arrived in Labrador around 1000. It is possible that Irish fishermen also knew of western lands. It has been argued that the Chinese admiral Zheng He reached the coast of North America in 1421. Archaeological evidence also points to settlers arriving on the west coast of the Americas from Polynesia and Japan. All of these travelers or explorers antedate Columbus. And indeed, Columbus died still thinking he had arrived at the Indies, in Asia. So it seems obvious that he is not the Old World discoverer of the Americas.

Let us grant the priority of all the alleged predecessors of Columbus. It still should remain an open question who the European or Asian discoverer of America was. For from the standpoint of community knowledge none of the earlier visitors managed to make the existence of two continents known to a broad audience. Even Columbus remained ignorant of the lands he had arrived at. But there is a crucial difference with Columbus: in his wake came other visitors who, a generation after his death, realized that the lands they came to were not part of the world known to Europeans or Asians. In 1507, fifteen years after Columbus' first visit and a year after his death, a map appeared identifying the new continents as the Americas. Balboa led an expedition of Spaniards who saw the Pacific Ocean for the first time in 1513. Magellan led a squadron of ships that circumnavigated the globe, establishing its rough dimensions, in 1519–1522. In some sense, then, the discovery of the New World extended over a period of at least thirty years, from 1492 to 1522, and the work of exploring, map-making, and charting continued long after the initial burst of exploration, conquest, and emigration. The expedition of Columbus set a process in motion that changed human knowledge and human interaction in an irreversible way. There is no doubt there were earlier contacts of some sort between Eurasians and Americans. But no one before Columbus forged a permanent link between the two worlds. After Columbus there were regular political, social, commercial, and intellectual exchanges between Eurasia and Africa on the one hand, and the Americas on the other.

Now there has been much discussion of the ravages visited upon the New World by visitors from the Old: conquests, epidemic diseases, slavery, and other forms of exploitation. And indeed when the Old World discovered the New the consequences were dire for the latter. Nonetheless, from the standpoint of knowledge, the meeting of the two worlds meant a great advance, for the inhabitants of both. Europeans discovered America; Americans discovered the Old World (some were taken there). There was a mutual discovery, and it resulted directly from Columbus' expedition, not from the travels of his Norse, Irish, or Chinese predecessors. Columbus is important not for being the first European or Asian in modern times to set foot on the New World, but for being the one whose visit changed the map of the world. His visit had social, political, military, geographical, historical, religious, ethnographical, biological, and other implications so profound that it changed the world irreversibly (not always for the better of course). Whether we praise or condemn Columbus (or do both in different contexts), we should recognize him as the European discoverer of the Americas, not for being the first European or Asian to land on western soil, but for setting a chain of events in motion that brought about a new level of mutual awareness between two worlds that had hitherto been, with few and limited exceptions, completely ignorant of each other's very existence. One might maintain that such a discovery was inevitable, given the facts of geography and the technological advances of the age. Nevertheless, it was a historical contingency who made the discovery and when, and which nation sponsored the voyage, and those contingent facts made a very large difference for the subsequent history of the world.

In the case of the discovery of America we end up justifying commonly held opinions rather than revisionist views, in a somewhat Aristotelian exercise.[2] Traditional answers to the question make sense once we realize that scientific discovery is part of process in which society comes to accept justified advances in knowledge. There is no guarantee that commonly held opinions will always triumph, and indeed in the case of Anaxagoras I shall argue for a revisionist view.

In any case, it is in this sense I wish to ask whether Anaxagoras was the discoverer of hitherto unknown astronomical relationships. Did he communicate astronomical knowledge to others in such a way that they could no longer be unaware of those relationships? Or did he arrive at this awareness and then fail to convince others of his

---

2. See *Nicomachean Ethics* 7.1.

success? Was he like a Columbus returning to Spain to face a skeptical audience? Herodotus tells the tale of a Phoenician expedition sent out by Pharaoh Necho (II) that sailed from Egypt across the Mediterranean, then south along the Atlantic coast of Africa until it had circumnavigated Africa after a three-year voyage.[3] The story had it that the sun was to their right or north part of the time. He does not believe this fantastic report, but its very implausibility to ancients and possibility to moderns convinces many modern scholars that the report was true. Yet this voyage seems not to have changed the world. Similarly, Aristarchus used sophisticated arguments to propose a heliocentric astronomy in the third century BC, but he failed to persuade almost anyone of his theory, so we do not regard him as the discoverer of the correct cosmography of the solar system. What remains to explore is the reception Anaxagoras' theory had among his contemporaries and successors. Typically each Presocratic theory won a few followers and failed to win over many others. Is that the fate of Anaxagoras' theory, or did he do better?

6.2 ANAXAGORAS' THEORY

First let us do what we have not hitherto done: to look at Anaxagoras' natural philosophy as a whole. We have already examined a few parts of the cosmology, but our main focus has been on his astronomy, and on some special features of that. Yet Anaxagoras did not present his astronomy in a vacuum; it was part of a general cosmology. So we must confront the cosmology as a whole. (He made important contributions in ontology as well, but those do not immediately bear on cosmological questions; I will deal with his ontology only cursorily.) We must remember, too, that we arrive at an understanding of his cosmology

---

3. Herodotus 4.42.

through a reconstruction of his theory. Although we get numerous reports of details of this theory, they come to us piecemeal, and even the summaries of Diogenes Laertius, Hippolytus, and others, consist of a later reconstruction from individual doctrines collected in the doxographical tradition. These summaries often do not tell us precisely why Anaxagoras held the views he did or how the several doctrines fit together. Yet it is possible to reconstruct his theory with some plausibility, and that is what we must do here, however briefly.

In the beginning,

> Together were all things, boundless both in quantity and in smallness.... And as all things were together nothing was manifest by reason of its smallness. For air and aether dominated all things, both being boundless. (B1)

Anaxagoras posits a large number[4] of elementary stuffs, which originally were completely mixed together in a primeval chaos. The complete mixture of everything in everything remains even now (B6), but different qualities emerge as different elements come to dominate in given regions. This results from the action of a vortex, revolving slowly at first, and then more rapidly (B12, B13, B9). The revolution was caused by mind (*nous*), which started the motion. Anaxagoras treats mind as like a physical body, except that it does not mix with the elements, which "would hinder it from ruling any object in the way it does when it is alone by itself" (B12).

> And the kinds of things that were to be—such as were but now are not, all that now are, and such as will be—all these did mind

---

4. They are *apeira* in quantity; but I hesitate to call them infinite in number, partly because the notion of infinity seems too mathematically precise at this point in history. After Anaxagoras and Zeno, and as a result of their speculations, the notion of infinity emerges, clearly in Aristotle *Physics* IV.

set in order, as well as this revolution, with which the stars, the sun, the moon, the air, and the aether which were being separated now revolve. (B12 in part)

Thus the revolution of the heavenly bodies is a product of the motion initiated by mind. As a result of the revolution

The dense, <the>[5] wet, the cold, and the dark came together here, where earth now is, while the rare, the hot, <the bright>,[6] and the dry retreated to the farther parts of the aether. (B15)

By the action of the vortex motion, earth came together into a heavenly body, while the lighter elements produced the upper atmosphere.

The earth is a large flat body, circular in shape, held in place by air pressure. Around it revolve the heavenly bodies. The sun is a "fiery molten mass," like a sphere of hot ore. The moon is earthy and illuminated primarily by the light of the sun, though, as we have seen, it has a secondary light which is visible at the time of a lunar eclipse. The stars are stony bodies that seem to be fiery; yet at least some of them are illuminated by the light of the sun. All the heavenly bodies, being heavy, are held aloft by the force of a powerful cosmic vortex. The heavenly bodies, according to Plutarch, "shine by resistance of the revolving aether."[7] This suggests that something like friction causes their fiery nature. Anaxagoras also holds that the stars travel in a cooler region than the sun, which, together with their greater distance, accounts for their inability to radiate warmth to the earth.[8] It suggests an alternative account of

---

5. Supplied by Diels.
6. Supplied by Schorn.
7. Plutarch *Lysander* 12.2.
8. Hippolytus *Refutation* 1.8.7.

the heat of the sun as well: perhaps the sun is hot because it travels in the aether, a fiery substance of the upper atmosphere. As we have seen, there are also asteroids circling lower than the moon and normally invisible to observers on earth. They can cause lunar eclipses and could be responsible for the phenomena we call comets. Apparently the asteroids are broken away from the earth at its edge by the force of the vortex; if they rise high enough, they become stars, which shine as a result of their motion in the upper portion of the heaven.[9]

Anaxagoras' astronomical advances are embedded in this cosmological theory. Heliophotism is accounted for by the fiery sun's illuminating the earthy, opaque moon. Antiphraxis accounts for eclipses, with the earth's or asteroids' blocking the sun's light to the moon, and the moon's blocking the sun's light to earth. The astronomical explanations could have been based on a different physical model, but do constrain whatever model is used. The sun does not have to be a stone lit by friction, but it does have to have its own source of light. The moon does not have to be earthy with mountains and valleys, but it does have to be spheroid, opaque, and consequently dense and massy. The sun and moon have to travel under the earth, so that the earth must be suspended in space and finite in size. The sun and the moon have to be permanent fixtures of the heavens, and so it is reasonable to think the stars are also permanent. The moon in particular must be massy, so it is reasonable to think all the heavenly bodies are so. Since they are massy, an adequate model of the heavens requires a strong force to hold the heavenly bodies in orbit; hence some kind of vortex theory is necessary. Thus there is a cohesiveness to Anaxagoras' cosmological theory, and the astronomical features he identifies put constraints on physical theories that are intended to account for them.

9. B16, Tigner 1979.

6.3 OTHER THEORIES OF THE FIFTH CENTURY

6.3.1 *Empedocles*

Empedocles shares a good deal with Anaxagoras, and since we have reason to believe Anaxagoras originated the theory, we can say that Empedocles learned a great deal from Anaxagoras. Empedocles seems to defend a complex cosmic cycle in which a differentiated world alternates with a Sphere in which all elements are unified into a single whole. This so-called cosmic cycle has become more controversial in recent times, with some scholars rejecting it altogether and some proposing new readings.[10] Further, Empedocles' cosmology must somehow be reconciled with his theory of reincarnation and divinity. His hexameter verse and very poetic diction make interpretation difficult at best. Yet fortunately his astronomy is largely independent of his cosmology, so I shall focus on the former while trying to steer clear of the problems of the latter.

Empedocles clearly accepts heliophotism, as several of his fragments indicate:

> Thus the [sun's] ray, having hit the broad circle of the moon... (B43)
>
> [The moon] gazes into the bright circle of her lord's face. (B47)
>
> [The moon] spins around the earth, a circular borrowed light. (B45)

The sun's rays illuminate the moon (B43); the moon looks on the face of "her lord," the dominant heavenly body (B47)—in other words, her

---

10. The classic study of the cosmic cycle is O'Brien 1969, which argues for two zoogonies in four stages. For simpler cycles with one zoogony, see Bollack 1965, vol. 1; Hölscher 1965; Solmsen 1965; Long 1974. On these works see Graham 1988. The discovery of new lines from the fragments has increased the material to work with; see Martin and Primavesi 1999. A new interpretation of the cycle is that of Sedley 2007, ch. 2.

luminous face is always turned towards the sun. The moon is a "circular borrowed light" *kukloteres... allotrion phōs* (B45), in a line that imitates Parmenides B14 and repeats his last two words (with its Homeric wordplay). The last fragment is important for showing that Empedocles recognized Parmenides as the source of heliophotism, which Empedocles fully understood. He was, as Theophrastus said, an enthusiast of Parmenides[11] and read Parmenides as proposing the sun as the source of the moon's light. Some scholars have claimed that Parmenides' statement is ambiguous, and not a secure starting point for heliophotism.[12] But at least one of his devotees saw him as making precisely the point Anaxagoras makes more clearly, and he embeds his own allusion to Parmenides in a clearly heliophotistic account of the moon.

Empedocles also gives an account like Anaxagoras' concerning the solar eclipse, as we have seen:

> .... [the moon] did away with his [the sun's] rays
> to the earth from above, and it obscured the earth
> as much as was the width of the bright-eyed moon. (B42)

The moon blocks the sun's rays to the earth in such a way that the shadow on the earth is equal to the diameter of the moon. This kind of shadow effect requires that the moon be significantly nearer the earth than the sun. Empedocles makes such an assumption:

> Empedocles says the moon is twice as far from the sun as from the earth. (Aetius 2.31.1 (ps.Plutarch) = A61)

Since the moon is relatively close to the earth, its shadow can approximate the size of the moon itself.

---

11. *Parmenidou zēlōtēs* Diogenes Laertius 8.55 = Theophrastus fr. 227 B = 28A1.
12. Tannery 1930:216–19; Guthrie 1962–1981, 2:66. To the contrary, see Coxon 1986:245.

The doxography reports a complicated theory of solar light, according to which there are two suns, the apparent one a reflection of the real sun, which is in the opposite hemisphere.[13] All of this seems confused and incoherent, for the report makes the apparent sun's light dependent on fire in the other hemisphere of heaven. Yet we have a fragment that clearly makes the sun the only cause of day and night:

> Earth produces night by obstructing the light [of the sun]. (B48)

So the opposite hemisphere from the one the sun is in will be dark, not bright, as the earth blocks the sun's rays to it. When the sun sinks below the horizon, it is night because the sun's rays no longer reach the earth's surface. The one and only sun is the one and only cause of daylight.[14]

Yet Empedocles does differ from Anaxagoras in his astrophysics (if we can trust the doxography on other points). The moon is not spherical but "lentoid," being a compression of air that is said to be cloudy or like hail (A60). Since Empedocles makes the moon lentoid, he flattens it but keeps enough convexity to allow for the shadows on its surface. He builds the moon of lighter-than-air materials unlike Anaxagoras. The sun is reported to be "ice-like," but its exact constituents are not named. As for the stars, we get the following reports:

> Empedocles [says the stars] are fiery bodies from fiery stuff, which the air [aether?] containing in itself extruded during the first segregation [of elements]. (Aetius 2.13.2 = A53)

> Empedocles [says] the fixed stars are attached to the icy sphere while the wandering stars [planets] are unattached. (Aetius 2.13.11 = A54)

---

13. Aetius 2.20.13, 2.21.2 = A56.
14. See Wright 1981:201–02; Kingsley 1994.

Empedocles holds that there is an ice-like sphere at the periphery of the world to which the fixed stars are attached. If the stars are composed of fire or fiery matter, they carry their own source of illumination. Since they are attached like nails to the crystalline sphere, we can understand why they keep their relative positions perfectly as they circle the earth. On the other hand, it is not clear how a fiery body can attach to an icy body without either the former being quenched or the latter melted. Possibly the heavenly sphere only looks like ice, but is not itself cold.

Insofar as the heavenly bodies are composed of light materials, Empedocles could get by with a less violent driving force for the heavenly rotation than a cosmic vortex. Yet he seems to accept the vortex, and to give it an earthly analogue:

> That is why all who make the heaven come to be say the earth is formed in the middle. They seek the reason for its staying in place, some claiming that width and size of it is the reason, others like Empedocles claiming the revolution of the heaven as it circles with a swifter motion prevents the motion of the earth just like water in a ladle. For when a ladle is swung in a circle, although the water is often suspended below the bronze, it does not travel downward with its natural motion for this same reason. (Aristotle *On the Heavens* 295a13–21 = 31A67)

Empedocles provides the experiment of swinging a ladle full of water around in a circle (with the concave part facing the person swinging it). The water is held in the ladle even when it is upside down as centrifugal force holds it to the container. So the heavenly bodies are whirled around the earth. The earth is held in place by the surrounding vortex. It has a finite size, as Empedocles claims, criticizing those who say it is boundless (B39), apparently in a criticism of Xenophanes' model of an infinite plane of earth, below which earth reaches down without end.

Thus Empedocles' cosmos is similar to Anaxagoras'. He resists a completely spherical moon (and probably sun), but keeps it spheroid. He constructs his heavenly bodies of lighter materials, but maintains the cosmic vortex. His earth is held in place by the surrounding vortex, while Anaxagoras' is supported by air pressure (but perhaps contained by a vortex also). Empedocles makes heliophotism a basic principle of his astronomy, poetically acknowledging a debt to Parmenides. He accepts antiphraxis as the cause of eclipses and of nighttime. If we can put Anaxagoras' work earlier than his—as I have argued in the previous chapter—we can say he is strongly influenced by Anaxagoras' astronomy and cosmology.

6.3.2 *Diogenes of Apollonia*

We do not know a great deal about Diogenes of Apollonia. He was active in the middle to the late fifth century. His physical theory was based on material monism, with air as the basic reality—a theory quite different from the pluralism of Anaxagoras. Yet his cosmology shows strong affinities with that of Anaxagoras. He held that the heavenly bodies were stony:

> Diogenes [says] the heavenly bodies are "like pumice-stones," and he thinks they are exhalations of the world; they are fiery.
> (Aetius 2.13.5 = 64A12(a))

The precise description of the kind of stones the heavenly bodies are like suggests a concrete model. Pumice is a volcanic rock, often of a burnt color with many holes in it. It is light because of the holes and may float on water. We are reminded of the meteorite of Aegospotami. Indeed, Diogenes made the connection himself:

> Invisible stones are carried around with the visible heavenly bodies, and because they are invisible they are unnamed; often they fall to earth and are quenched like the meteor of stone that fell in flames on Aegospotami. (Aetius 2.13.9 = 64A12(b))

Like Anaxagoras, Diogenes holds that there are asteroids, which can fall out of orbit to become meteors. Obviously the meteorite of Aegospotami casts a long shadow in fifth century natural philosophy.

How precisely he conceives of astronomical events is less clear. We get the following reports:

> Diogenes [says] the sun is "like a pumice-stone," to which rays from the aether are attracted [?]. (Aetius 2.20.10 = 64A13)
>
> Diogenes [says] the sun is quenched by cold colliding with its heat. (Aetius 2.23.4 = 64A13)
>
> Diogenes [says the moon] is a like an "ignited pumice-stone." (Aetius 2.25.10 = 64A14)
>
> Diogenes [says] the comets are stars. (Aetius 3.2.8 = 64A15)

A13 and A14 reiterate the point of A12(a) that the heavenly bodies are like pumice stones. It adds a source for the sun's heat, some kind of connection to the rays of the aether, the clear upper air that is often thought to have a fiery character. Presumably, then, the sun is hot not from friction but from some kind of interaction with a fiery stuff. It is quenched by cold. When precisely is this meant to happen? On setting, or during an eclipse? In fact the opinion is recorded in a chapter on the "turnings" or solstices of the sun (Aetius 2.23; v. 4 Diels, v. 6 Mansfeld and Runia). Thus perhaps "quenched" is not the right translation (though it is the usual one) but something like "dampened" or, in the precise context, "turned back." In any case the opinion seems to come from a discussion of solstices rather than daily settings.[15]

Diogenes holds that the moon is ignited. This may suggest having its own source of light. But this is consistent with Anaxagoras' theory according to which the moon has a secondary source of light, even though its primary source is the sun. The comets could be stars that are more luminous than asteroids but less so than the planets and fixed stars.[16]

Diogenes' theory could, on the other hand, be incompatible with that of Anaxagoras. But nothing in the record requires this. The fact that Diogenes allows for asteroids like Anaxagoras (64A12(b)) suggests that he accepts the latter's motivation for positing them: they allow for lunar eclipses in some problematic cases. Unfortunately we do not have enough of his background theory to see how the pieces fit together. But what is clear is that the meteor of Aegospotami has become a datum that cannot be ignored in theory construction, and that Anaxagoras' general picture wields great influence with Diogenes. He accepts the Lithic Model as a starting point of his theorizing.

6.3.3 Philolaus

Writing in the last third of the fifth century, Philolaus produced the most innovative astronomy of the Presocratic period. The only Presocratic to depart from a geocentric theory, he created what we might call a hestiocentric theory: the center of the world was *hestia*, the hearth of the universe, a central fire.[17] Around this revolved ten heavenly bodies, including the earth, which he saw for the first time as a planet rather than as a unique central body:

---

15. Thus Mansfeld and Runia 2009:560–61; to the contrary Laks 2008:206–08.
16. Laks 2008:204, 209.
17. Maniatis 2009 calls the universe "pyrocentric," but I prefer my term because whereas fire is not localized in one place in the universe, the hearth is.

And the sun too, since it itself is supposed to be responsible for the times and seasons, according to [Aristotle], they say is located there where the number seven is, which they call "timeliness." For it holds the seventh place of the ten bodies moving around the center and hearth, and it is in motion after [i.e., below] the sphere of the fixed stars and the five spheres of the wandering stars; after which the moon is eighth, the earth ninth, after which is the counter-earth. (Alexander *Metaphysics* 38.20–26, Aristotle fr. 203; cp. Aetius 2.7.7 = A16)

Philolaus makes an important advance just by recognizing the five planets. No earlier Greek system had identified five planets, as far as we know. And indeed, Eudemus (the Peripatetic investigator of early astronomy) acknowledges the Pythagoreans to be the first to recognize the proper order of the five planets:[18]

There [in studies of astronomy] it is demonstrated concerning the order of the wandering stars, their size and distances that Anaximander was the first to provide an account of their sizes and distances, as Eudemus reports, attributing to the Pythagoreans the first determination of the correct order of their positions. (Simplicius *On the Heavens* 471.2–6, Eudemus fr. 146)

By allowing the earth to make one revolution per day around the central hearth, Philolaus can account for the cycles of day and night. The sun seems to reflect the light to the earth:

Philolaus the Pythagorean [says] the sun is glassy, receiving the reflection of the fire in the world and filtering through to us its

---

18. There is no corroboration for the claim that Anaximander even distinguished planets from fixed stars. See above, ch. 2.3, ch. 3.2.

light and heat, so that in a sense there are two suns: the fiery one in the heaven and the one that is fiery by mirroring it—unless one should call the rays dispersed to us from the reflection of the mirror a third sun. For we call the last 'sun' as being the image of an image. (Aetius 2.20.12 = A19)

This seems a bit confused and confusing, but it makes sense if we take "the fiery [sun] in the heavens" as an outer fire: then the job of the glassy sun that orbits the central fire is to disperse that fire to the outer parts of the world. Incidentally, this report suggests a possible source for Empedocles' allegedly lens-like sun: perhaps doxographers confused Empedocles with Philolaus.

We get one detailed report of Philolaus' views on the moon:

> Some of the Pythagoreans, including Philolaus, [say] its earthy appearance arises from the moon's being inhabited, just as is our earth, by animals and plants, but larger and more beautiful than ours. For they are fifteen times more powerful than animals here, and do not make excretions, and a day there is that much longer than here. (Aetius 2.30.1 = A20)

This is a distinctly curious account, with a quality of science fiction about it. Yet we can recall that Anaxagoras believed that "the moon has dwellings" (Diogenes Laertius 2.8). Evidently Philolaus accepts his account of the moon as a solid, earthy body that is not different in kind from earth but rather in degree of livability. A day on the moon will be, on the model of Philolaus as on that of Anaxagoras, about fifteen days long (half a lunar month), which suggests the number associated with animals.[19]

---

19. There is, however, a problem, in that the fifteen-day lunar day should properly compared with the twelve-hour daytime on earth, which being half a (solar) day long, makes the lunar day thirty times longer than an earthly day. See Huffman 1993:276.

As to Philolaus' views on heliophotism and antiphraxis, we are not told directly. Yet there are some interesting views attributed to Pythagoras or the Pythagoreans. Aetius reports that after Thales, Pythagoras, Parmenides, Empedocles, Anaxagoras, and Metrodorus thought the moon was lit by sunlight.[20] Since Pythagoras left no writings, we do not know what his views on this subject were. But often when Aristotle reports Pythagorean views of cosmology and astronomy, his reports portray those of Philolaus, as we can tell from peculiarities of his system, including the counter-earth.[21] "Pythagoras" may well be a place-holder for the earliest Pythagorean cosmology, which is that of Philolaus.

There is an intriguing report of a Pythagorean view concerning eclipses of the moon:

> According to the research of Aristotle and the statement of Philip of Opus, some of the Pythagoreans [say eclipses of the moon happen] by reflection and obstruction (*epiphraxis*), sometimes by the earth, sometimes by the counter-earth.
>
> There are some of the younger Pythagoreans who thought that [it happened] when the fire that was kindled spread gradually in order until it produced a complete full moon, and in turn it decreased correspondingly until the conjunction, at which time it was completely quenched. (Aetius 2.29.4, vv. 4–5 Mansfeld and Runia)

The first sentence of this account explains lunar eclipses by "reflection and obstruction," that is, they say that eclipses happen as a result of something obstructing the sun's light, which is normally reflected from the moon. There are two sources of the obstruction: the earth

---

20. Aetius 2.28.5.
21. Philolaus' cosmological system "remains the only complete system that can be assigned to the early Pythagoreans," although there were other explanations of particular phenomena (Huffman 1993:239).

and the counter-earth. Since only Philolaus, to the best of our knowledge, posits a counter-earth, this must be his theory. He too accepts the antiphraxis theory, with the addition that he thinks the counter-earth can cause an eclipse.

Some have questioned the possibility of the counter-earth's playing any role here.[22] But in fact Philolaus needs it. On this theory, the earth revolves around the central fire, with the moon's orbit outside the earth's and the sun's outside the moon's. The earth presumably is spherical, as theorized by Parmenides (though we receive no confirmation of this from our sources).[23] The side of the earth we are on must always point away from the central fire, or we would see it. Now when the moon suffers an eclipse at its zenith, Philoaus can perfectly well explain the darkening of the moon as caused by the interposition of the earth between it and the sun (fig. 6.1). But when the moon suffers an eclipse on rising or setting, he must suppose that the sun, moon, and earth are not lined up, since the sun and moon are in opposition, but the earth does not then stand in a line with the sun and moon (see figure 6.2). Thus we need something else to block the sun's rays, and the counter-earth can fill this function. (Philolaus seems to reject asteroids because of his interest in a simple numerical scheme of ten heavenly bodies.) The counter-earth is needed to account for lunar eclipses that happen at the rising or setting of the moon (see figure 6.3).

The second sentence of Aetius 2.29.4 (v. 5 Mansfeld and Runia) follows the first in Stobaeus. There is no guarantee that the "younger Pythagoreans" include Philolaus; he is younger than several figures in the school, but older than the generation of the fourth century. In any case, the most important feature of this account is that it seems confused: what the sentence describes is not a lunar eclipse—which

---

22. Huffman 1993:246–47.
23. Burch 1954 has the counter-earth traveling along the earth's orbit opposite it, and being, like the earth, disk-shaped. According to him, its mass balances the earth's, producing equilibrium in the cosmos.

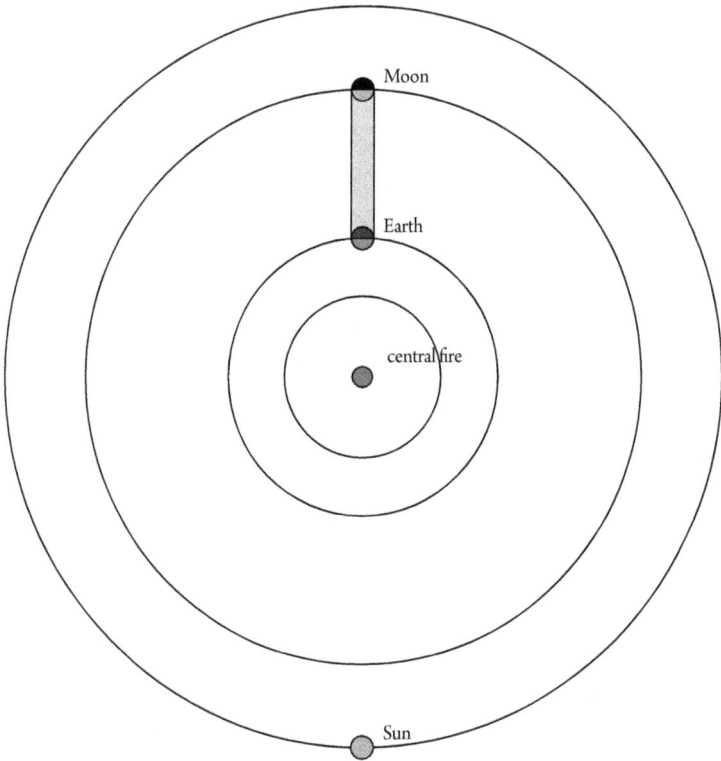

Figure 6.1. Philolaus: Nocturnal Lunar Eclipse. Earth blocks sun's light to moon.

happens in hours, not in the course of a month—but rather the phases of the moon. It is odd that the increase is described in terms of fire rather than light, but otherwise the account gives a straightforward description of heliophotism. So some younger Pythagoreans accepted heliophotism. And since the antiphraxis theory of Philolaus presupposes heliophotism, he must be one of those who would accept the account, whether Aetius (and his sources) intends to include him or not.

Thus Philolaus subscribes to Anaxagoras' general principles of lunar light and eclipses, even though his cosmology and general astronomical model are radically different from those of his

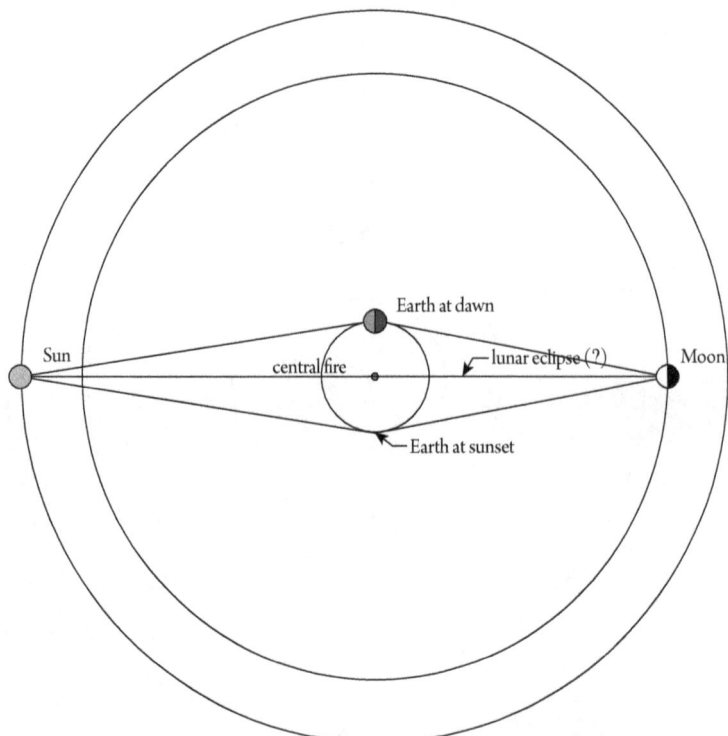

Figure 6.2. Philolaus: Crepuscular Eclipse. Earth is out of position to block the sun's light to the moon.

predecessor. The power of heliophotism and antiphraxis is not lost on the first Pythagorean astronomer.

6.3.4 Democritus

Democritus acknowledges Anaxagoras as a predecessor,[24] and he clearly gets a great deal of his cosmology from him. For instance, his theories of the Milky Way and of comets are the same as those of

---

24. Diogenes Laertius 9.34 = 68A1.

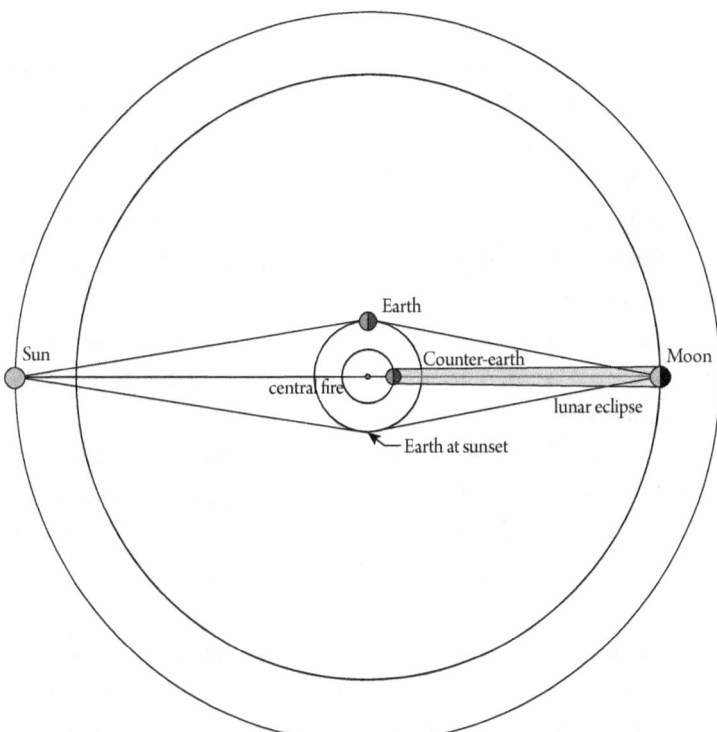

Figure 6.3. Philolaus: Crepuscular Eclipse Reconstructed. Counter-earth blocks the sun's light to the moon.

Anaxagoras.[25] At the same time he is said to have accused Anaxagoras of plagiarizing his theory of the sun and the moon from more ancient sources.[26] Whether he has in mind Thales or some unnamed sources such as suggested by Plato in the *Cratylus* we are not told.[27] Yet he

---

25. Aristotle *Meteorology* 342b27–29, Aetius 3.2.2 = 59A81, cf. 68A92; Aristotle *Meteorology* 345a25–31 = 59A80, Olympiodorus *Meteorology* 67.32–37, cf. 68A91.
26. Diogenes Laertius 9.34, cf. 9.35. "Despite his avowed dislike of Anaxagoras, whom he accused of plagiarizing old ideas about the sun and moon (B5), many of Democritus' own astronomical ideas ... are identical with those of Anaxagoras" (Dicks 1970:82).
27. *Cratylus* 409a–b.

seems to have followed Anaxagoras in his general account of the sun and moon. He holds that they are made of earthy materials and held in orbit by a vortex motion. When the circle about the sun expanded, fire was enclosed in it—whatever that means.[28] Aetius reports that the sun is a "fiery molten rock" as in Anaxagoras.[29] The moon has mountains and valleys, as in Anaxagoras, and its mottled appearances is caused by its topography and the resulting shadows.[30] The reason the moon seems to move eastward relative to the stars faster than the sun is that it is lower and receives less impetus from the cosmic vortex in its westward whirl.[31] He clearly accepts the Lithic Model as a basis for cosmology.

Unfortunately, we have little record of Democritus' views on lunar light and eclipses. His predecessor Leucippus expressed some view on eclipses that was lost in a lacuna.[32] We get some hint from Plutarch that Democritus accepted the view of heliophotism:

> "But," says Democritus, "[the moon] stands in line with its illuminator [at conjunction] and accepts and receives the sun." So it is reasonable that the moon should be visible and the sun's light shine through it. (*The Face on the Moon* 929c = A89a)

This looks at first glance like an argument for why the moon should be translucent, but it comes from a speaker in a dialogue making an objection to the view that the moon is ice-like (929b), a view that the speaker does not actually attribute to Democritus—he rather borrows Democritus' words to describe the relative positions of the sun

---

28. ps.Plutarch *Miscellanies* 7 = 68A39.
29. Aetius 2.20.7 = A87.
30. Aetius 2.30.3 = A90.
31. Lucretius 5.621–36 = A88.
32. Diogenes Laertius 9.33.

and moon and the moon's reception of light from the sun.[33] The point of the objection is that the moon cannot be translucent, as Anaxagoras had already recognized. Democritus clearly regards the sun as the moon's "illuminator" (*phōtizōn*), and hence accepts heliophotism. Given the heavy influence of Anaxagoras on Democritus' astronomy and cosmology in general, it is likely the latter understands the moon as an opaque, earthy body, and he may accept antiphraxis as well.[34]

6.4 CHARACTERISTICS OF THE LITHIC MODEL

In ch. 2.3 above, I went through characteristics of sixth-century cosmologies, which I grouped together as exemplifying the Meteorological Model. We are now in a position to compare and contrast those with fifth-century cosmologies. I will proceed in the same order as for the Meteorological Model.

1. The theories of the fifth century predominantly envisage a flat earth, but at least Parmenides and Philoaus envisage a spherical earth.
2. Theories of the Lithic Model allow for heavenly bodies to move below as well as around and above the earth.
3. Besides stars, sun, and moon, theories of the fifth century recognize planets and comets, and sometimes posit the existence of asteroids and meteoroids.

---

33. Thus the close-quote should come where I have marked it, as in Cherniss 1957:103, rather than at the end of the next sentence, as in Bailey 1928:151. There does seem to be one philosopher who views the moon as partly translucent: Ion of Chios, according to Aetius 2.25.11 = 36A7: "Ion (declares that it [the moon] is) a body that is partly glass-like and transparent, partly opaque [*aphenges*]" (trans. Mansfeld and Runia, v. 12 in their edition).
34. Görgemanns 1970:36n.69 takes the passage as accounting for solar eclipses; but that context is not evident from the citation.

4. The heavenly bodies are strongly distinguished from meteorological phenomena. They move in a realm farther out than the immediate atmosphere of the earth and are subject to different influences and forces.
5. The heavenly bodies are understood to be permanent rather than ephemeral bodies. They are not necessarily everlasting—they arise through cosmogonical processes, and some of them may fall out of orbit and onto the earth, as in Anaxagoras' theory, and some stony bodies may be torn away from earth's surface to join the heavenly host. But most of them are permanent fixtures of the heavens which last for long periods of time.
6. Heavenly bodies are heavier than air, of a predominantly earthy or stony composition. Some of them may fall to earth to become meteorites.
7. Some heavenly bodies have their own source of light, most notably the sun, while some are dark and depend for their light on the sun. There may be bodies such as the moon which have both a dim native illumination and receive the sun's rays. There may be some bodies that under normal conditions are dark and invisible but by contact or some heating process become fiery, as with meteors. The light of heavenly bodies is caused not by combustion of fuel but by some other process, either friction or the attraction of aether or their movement in a fiery realm of the heaven.
8. Heavenly bodies are spherical or spheroid in all cases.
9. The motion of heavenly bodies is caused by a powerful cosmic vortex motion that keeps heavy bodies suspended. The different speeds of heavenly bodies may be indicative of their altitude above the earth and the power of the vortex at that altitude. Their speed is not to be compared with that of objects in the vicinity of earth.

Thus we come to the following points of comparison.

Table 6.1 EARLY GREEK COSMOLOGIES

Features	Sixth-Century Cosmologies	Fifth-Century Cosmologies
Earth	flat	flat or spherical
Region	bodies above, around the earth	bodies above, below, around the earth
Entities	sun, moon, stars	sun, moon, stars, planets, comets, asteroids
Realm	continuous with meteorology	distinct from meteorology
Duration	ephemeral	permanent
Weight	light (fiery or airy)	heavy (earthy or stony)
Source of Illumination	self-luminous, by combustion	self-luminous and non-self-luminous, not by combustion
Shape	various	spherical or spheroid
Cause of Motion	wind, atmospheric conditions, residual motion	vortex

It should be evident at this point that there was a significant development in cosmography from the sixth to the fifth century. The development is consistent with the discovery of heliophotism and antiphraxis and, if the argument of this book is right, is mainly determined by the acceptance of those new theories. The fact that fifth-century cosmology is so closely linked to the theory of Anaxagoras provides historical evidence for the influence of the conceptions he and Parmenides introduced. The major theories of the fifth century are instances of the Lithic Model, characterized by massy bodies whose behavior is independent of the lower atmosphere with its meteorological phenomena (see ch. 4.4 above).

6.5 THE DOXOGRAPHY

Ancient students of philosophy organized opinions of philosophers, as if they were unit-ideas or doctrines, into lists on given topics so as to survey all possible theories. They organized them dialectically, by a system of divisions as sketched by Aristotle (using a method that perhaps antedates him). They were laid out for analytical, not historical, purposes, and in a kind of logical rather than chronological order, but often they provide the only bits of information we have about the theories of early Greek philosophers. The simplest way to study the impact of Anaxagoras' theories is to follow them in the doxography. A succession of doxographers condensed the information from Theophrastus' reports, adding later figures to the lists of opinions. The most extensive collection of doxography is that of Aetius, compiled around the first century AD. Aetius' work was reconstructed by Diels (1879) from later compilations, and is now being reworked by Mansfeld and Runia.[35]

6.5.1 The Moon's Light

On the moon's light (book II, chapter 28) we get the following:

---

35. The classic work is Diels 1879, revisited in Mansfeld and Runia 1997. A recent study of Book II of Aetius' collection of opinions provides an up-to-date conspectus of his survey: Mansfeld and Runia 2009. In their analyses of the chapters they bring out the divisions that underlie the lists of opinions.

1. Anaximander, Xenophanes and Berosus (declare that) the moon has its own light [*idion . . . echein phōs*].
2. Aristotle (declares that it has) its own (light), but it is dimmer somehow.
3. The Stoics (declare that its light is) murky in appearance, for it is airlike.
4. Antiphon (declares that) the moon has its own gleam, and that the part of it that is hidden is dimmed by the approach of the sun, since it is natural for the stronger fire to make the weaker one dim, which indeed also occurs in the case of the other heavenly bodies.
5. Thales was the first to say that it is illuminated by the sun [*hupo tou hēliou phōtizesthai*].
6. Pythagoras, Parmenides, Empedocles, Anaxagoras, and Metrodorus (declare) likewise.
7. Heraclitus (declares that) the sun and the moon undergo the same experience: since they are heavenly bodies that are bowl-like in their shapes and receive their radiance from the moist exhalation, they light up in their appearance (to us), the sun doing so more brightly because it moves in air that is purer, whereas the moon moves in murkier (air) and for this reason appears dimmer. (trans. Mansfeld and Runia)

If we order the results of this survey chronologically, we get the following results, using "H" to designate "heliophotism," "I" to designate "idiophotism" (having its own light):

Thales	H
Anaximander	I
Pythagoras	H
Xenophanes	I
Heraclitus	I

Parmenides	H
Anaxagoras	H
Empedocles	H
Antiphon	I
Metrodorus	H
Aristotle	I
Berosus	I
The Stoics	I?

From this no pattern of dominance emerges. Heliophotism is the earliest theory, which wins some support, but, except perhaps for the early fifth century (Parmenides, Anaxagoras, Empedocles) wins over a majority of thinkers, while idiophotism seems to prevail in the long run.

But let us take a critical look at the list and the ascriptions it makes. In the first place there is no solid evidence to put Thales anywhere on the list. Aristotle seems to have no concrete information about his theories, other than his claim that all is water. It appears likely that Hellenistic scientists assumed from his alleged prediction of the solar eclipse that he understood eclipses. They consequently attributed to him the knowledge *they* then had of eclipses, which presupposed heliophotism. Thus the view that he espoused heliophotism is a backward projection of their own understanding as an ex post facto explanation of his eclipse prediction.

For Pythagoras we have no independent confirmation of his view. In many cases it appears that the first and only detailed Pythagorean astronomy was that of Philolaus. He did accept heliophotism, as we have seen. Again, it may have been assumed that since a Pythagorean scientist/philosopher accepted heliophotism, he must have gotten the doctrine from Pythagoras. Yet that would be a historically hazardous inference, given that, as far as we know, no sixth-century thinker understood heliophotism.

Thus we must remove Pythagoras from the list of advocates of heliophotism.[36]

We have seen reasons for accepting Anaximander, Xenophanes, and Heraclitus as advocates of idiophotism. We have also seen that Parmenides, Anaxagoras, and Empedocles accept heliophotism. What of Antiphon's alleged idiophotism? If we look at Aetius' report in verse 4, we see reasons to doubt his interpretation. What Antiphon seems to be saying is that the sun's light overpowers the moon's own natural light. In other words, the sun's light obscures the moon's own radiation when the former is falling on the moon's surface. We have here an account of the secondary light of the moon, much like that of Anaxagoras himself, and perhaps indebted to the earlier philosopher. What Antiphon is saying is that in addition to the reflected solar light, there is a glow emanating from the moon itself. On this account the phases of the moon are still caused by reflected solar light, while the glow visible in a lunar eclipse, and perhaps in earthshine, are the result of the moon's own radiation.[37]

As for Aristotle, Aetius is just plain wrong: Aristotle clearly accepts heliophotism and accounts for the phases of the moon as shadows generated by the sun's light.[38] He also accepts antiphraxis, as we shall see later, a theory that presupposes heliophotism. Berosus (or Berossus), a Babylonian priest who became Hellenized and came to teach in the Greek world at Cos around 300 BC, did apparently present a theory in which the moon had its own fire on half of its

---

36. Burkert 1972:299–322. "[T]here is no good reason to assume a mysterious, secret pre-Philolaus astronomy, belonging to Pythagoras or the Pythagoreans..." (317).
37. Aetius includes Antiphon's account among mixed views, that is, those that attribute both idiophotism and heliophotism to the moon: Mansfeld and Runia 2009:605–06, 608–09. Aetius 2.29.3 includes Antiphon among those who explain lunar eclipses by the bowl model used by Heraclitus. This is inconsistent with the account of the moon's light in 2.28. Pendrick 2002:297–98 (and cf. 295–97) accepts the bowl theory.
38. It is also inconsistent with Aetius' report of Aristotle in 2.29.6 (verse 7 Mansfeld and Runia), which implies heliophotism.

sphere.[39] But he structured his theory in such a way that the fiery part of the moon is always facing the sun—so that his theory makes no predictions that are at variance with heliophotism. No Greek thinker seems ever to have accepted his explanation, which in turn seems to offer no theoretical advantages to heliophotism, and to allow no independent empirical confirmation. On the other hand, Berosus' own theory seems dependent on heliophotism for its conception of how the moon relates to the sun, while not obviously improving on it.[40]

The doctrine ascribed to the Stoics in verse 3 tells us that the moon's light is dim or "murky" because it (presumable the moon) is "airlike." This is more a remark about the quality of the moon's light than one about its source, although it could possibly imply that the airy moon generates its own light. Yet, as we shall see presently, Aetius himself (along with others) attributes antiphraxis to the Stoics, which he explicitly links to heliophotism, in his next chapter. So whatever the Stoics are saying about the moon, they cannot be advocating idiophotism.

Let us then revisit the list in light of the corrections we have made, and adding Anaximenes to the list, whose views we can gather from other sources:

	Aetius	Corrected
Thales	H	?
Anaximenes	?	I
Anaximander	I	I
Pythagoras	H	?
Xenophanes	I	I
Heraclitus	I	I

39. Cleomedes 2.4.1–9; Vitruvius 9.2.1, 9.6.2; Lucretius 5.720–28. He also wrote a history of Babylonia; see Jacoby 1957–1963, # 680.
40. See Toulmin 1967. As Cleomedes points out, Berosus' theory cannot account for the darkening of the moon in lunar eclipses (*Caelestia* 2.4.12–17).

	Aetius	Corrected
Parmenides	H	H
Anaxagoras	H	H
Empedocles	H	H
Antiphon	I	H
Metrodorus	H	H
Aristotle	I	H
Berosus	I	I
The Stoics	I?	H

We get idiophotism by all thinkers before Parmenides, and heliophotism by all after, with the exception of the Babylonian expatriate, Berosus, coming from a Mesopotamian tradition. Within the Greek tradition itself, we have a remarkable and almost unparalleled consensus. Heliophotism carries the day and becomes the standard theory of lunar light from Parmenides on.

6.5.2 Theories of Solar Eclipses

Aetius treats of solar eclipses in chapter 24:

1. Thales was the first to say that the sun undergoes an eclipse when the moon with its earthy nature courses perpendicularly in between (it and the earth)....
2. The Pythagoreans and Empedocles hold a similar view.
3. Anaximander (declares that the sun is eclipsed) when the mouth through which the outpouring of fire occurs is blocked.
4. Heraclitus (declares that it undergoes an eclipse) in accordance with the turning of its bowl-like shape, so that the hollow aspect faces upwards and the convex aspect faces downwards in the direction of our vision.

5. Xenophanes (declares that it undergoes an eclipse) through quenching. But another sun occurs in the east....
6. Some (thinkers) declare that it is a condensation of clouds invisibly passing in front of the (sun's) disk.
7. Aristarchus makes the sun stand still together with the fixed stars, while he moves the earth in the circle of the sun and (declares that) its disk is obscured in accordance with the tiltings of this body (i.e., the earth).
8. Xenophanes declares that there are many suns and moons in accordance with the latitudes of the earth and its sections and zones. But at a certain moment the (sun's) disk falls into a section of the earth that is not inhabited by us, and in this way, as if stepping into the void, it appears to undergo an eclipse. The same (thinker declares that) the sun advances indefinitely, but seems to go in a circle because of the remove (away from us). (trans. Mansfeld and Runia)

Again, Thales is problematic in verse 1. In verse 2 we are not told who the Pythagoreans are; we may suspect Philolaus again because of his views on lunar eclipses, though we have no independent confirmation of his views on solar eclipses. The unnamed thinker(s) in verse 6 may be identified with Anaximenes, according to one attractive proposal.[41] Xenophanes' view is stated in verse 5 and expanded in verse 8. The sun seems to suffer eclipse when it travels into a desert area in which there is not enough exhalation or evaporation from earth to fuel its fire. Then it is extinguished temporarily. Oddly, we get no recorded mention of Anaxagoras, who clearly holds the antiphraxis view. From Aetius it is not clear what Aristarchus' precise view is. That he accepts antiphraxis for solar eclipses is seen clearly in verse 8

---

41. See above, ch. 2.2.3 and ch. 2 n. 79. If this is the view of Anaximenes, the name(s) attached to the opinion must have been lost in transmission.

of his treatise; that he accepts antiphraxis for lunar eclipses (not mentioned by Aetius) is seen in verses 13, 14, and 15.[42] Parmenides does not appear in this chapter or in chapter 29, as he had apparently said nothing about eclipses. Aristotle is not mentioned in chapter 24, but he clearly accepts antiphraxis.

Thus we get the following results, with "A" for "antiphraxis," "O" for any other view (the views of figures not mentioned by Aetius here are put in parentheses):

Philosopher	Solar Eclipses
Thales	?
Anaximander	O
Anaximenes?	O
Xenophanes	O
Heraclitus	O
Anaxagoras	(A)
Empedocles	A
Philolaus	A?
Antiphon	?
Aristotle	(A)
Aristarchus	A

Here again we see a pattern emerge. From Anaxagoras on, every philosopher or astronomer we know about accepts antiphraxis.

We might add that the Epicureans withhold judgment about the causes of eclipses. But even Lucretius seems to give pride of place to the antiphraxis theory.[43] It is part of the general strategy of the Epicureans not to commit themselves on theories about phenomena

---

42. See Heath 1913:383, 392–409. The starting point of Aristarchus' deductions is Hypothesis 1, "That the moon receives its light from the sun" (trans. Heath, 353), i.e. heliophotism.
43. 5.753–57. Cf. Epicurus *To Pythocles* 96.

remote from us.[44] Thus they are in principle averse to making pronouncements about astronomical explanations. Their reticence does not reflect so much the ambiguity of the evidence as Epicurean priorities whereby ethics trumps physics, and peace of mind is more important than scientific research.

Claudius Ptolemy, too, accepts the antiphraxis theory and hence transmits it to later astronomers and to the medieval and early modern world.[45] It appears then that almost every theorist accepts heliophotism and antiphraxis from the time of Anaxagoras on. To be sure, many laymen still believed in mythological accounts. But even educated non-philosophers came to accept the philosophical-scientific theories. Writing at the end of the fifth century Thucydides recognizes the astronomical conditions that make a solar eclipse possible, while he faults Nicias for his superstitious reaction to a lunar eclipse—which brought disaster to his army.[46] Thus the new astronomy began to spread even among the educated laity. Among experts, these two related theories emerge as not just interesting speculations, but foundational principles accepted by cosmologists and astronomers for ever after. It does not even matter whether one's cosmology is geocentric, hestiocentric (Philolaus), or heliocentric (Aristarchus); heliophotism and antiphraxis provide the only empirically supported theories of lunar light and eclipses. They win out in antiquity and pass on to the modern era.

6.5.3 Theories of Lunar Eclipses

In ch. 29, Aetius surveys theories of lunar eclipses:

---

44. Epicurus *To Pythocles* 85–87; *To Herodotus* 79–80.
45. *Almagest* 4.1, 267 Heiberg; 6.1, 461. See epigraph to this chapter.
46. Thucydides 2.28.1, cf. 4.52.1; on Nicias and the eclipse 7.50.4: he was "too devoted to divination."

1. Anaximander (declares the moon is eclipsed) when the mouth on the wheel (of fire) is obstructed.
2. Berosus (declares that it is eclipsed) in accordance with the turning of the uninflamed part (of the moon) towards us.
3. Alcmaeon, Heraclitus, and Antiphon (declare that it is eclipsed) in accordance with the turning of the bowl-like (shape of the moon) and its lateral motions.
4. Some of the Pythagoreans according to the research of Aristotle and the assertion of Philip of Opus (declare that it is eclipsed) through reflection and obstruction sometimes of the earth and sometimes of the counter-earth.
5. ...[46a]
6. Xenophanes (declares that) the monthly concealment too (takes place) by quenching.
7. Thales, Anaxagoras, Plato, Aristotle, the Stoics, and the astronomers (*hoi mathēmatikoi*) agree in unison that it (the moon) produces monthly concealments by travelling together with the sun and being illuminated by it, whereas it produces the eclipses by descending into the shadow of the earth which interposes itself between the two heavenly bodies, or rather when the moon is obstructed (by the earth).
8. Anaxagoras, as Theophrastus states, (declares that it is eclipsed) also when it happens that bodies (in the space) below the moon interpose themselves. (trans. Mansfeld and Runia)

Apparently Berosus uses the same mechanism for a lunar eclipse as for the new moon: the dark side of the moon is turned towards us. In the phases of the moon the rotation of the satellite takes a (lunar) month; in the case of an eclipse it must be rapid. Yet on his account, we should see a rapid reiteration of the phases of the moon, which we do not. Furthermore, on this view the fact that lunar eclipses happen

---

[46a]. This verse discusses views of some younger Pythagoreans; but it confuses lunar eclipses with heliophotism. See above, pp. 195–97.

at the time of the full moon is purely coincidental. The account of bowls in verse 3 also calls for a rapid turn of the moon. In the case of Antiphon, it is problematic how this account could be squared with the observation that the moon has its own secondary light (in addition to the sun's illumination), as described above in section 6.5.1. The Pythagoreans mentioned in verse 4 must consist of Philolaus, the only known theorist to posit a counter-earth.

In verse 7 we get the antiphraxis theory, with a large number of adherents. We must question Thales in accordance with principles mentioned concerning chapter 28. Plato does not account for eclipses in the extant corpus (which seems to be complete), so his mention is problematic. The other adherents seem to be correctly named. Empedocles too holds this view, though he is not named. In verse 8 we get a notice of Anaxagoras' other source of lunar eclipses, which we have found confirmed by other reports.

The Stoics from Zeno on clearly accept both heliophotism and antiphraxis:

> [The Stoics] believe that both the heavenly bodies and the earth (which is motionless) are spherical. The moon does not have its own light (*ouk idion echein phōs*) but takes it from the sun's illumination. The sun is eclipsed when the moon occults the part facing us, as Zeno illustrates (*anagraphei*) in *On the Whole*. For it is visible passing under and concealing the sun and then moving on. (This can be observed reflected in a basin full of water.) The moon is eclipsed when it enters into the shadow of the earth. Hence it is eclipsed only at the time of the full moon.... (Diogenes Laertius 7.145–46, *Stoicorum veterum fragmenta* I.119, II.650)

Diogenes goes on to give a description of the geometry of a lunar eclipse.[47] Before this passage he gives some details of Posidonius on

---

47. Diogenes reproduces a theory of lunar motions without understanding it: Edelstein and Kidd 1972–1988, 2:480–482.

the nourishment of the sun and moon, while other sources make it clear that Posidonius thought the sun's rays penetrated some distance into the moon, but in all of this Posidonius must accept the overall theory of Zeno, which is just Anaxagoras' theory with added precision.[48] Diogenes' report suggests that Zeno used diagrams and gave advice on the safe viewing of a solar eclipse.

Verse 7 mentions also the astronomers, that is, the mathematical astronomers. The reference is vague, but it seems to apply to virtually all mathematical astronomers. We have seen that Aristarchus and Ptolemy accept heliophotism and antiphraxis. The great Hipparchus clearly accepts heliophotism as a starting point of his calculations of the relative sizes of the earth, sun, and moon.[49]

Thus we come to the following result:

	Solar Eclipses	Lunar Eclipses
Philosopher		
Thales	?	?
Anaximander	O	O
Anaximenes?	O	
Xenophanes	O	O
Alcmaeon	O	

---

48. Cleomedes *Caelestia* 2.4.105 = Posidonius fr. 123 Edelstein-Kidd; Plutarch *The Face on the Moon* 929d, 932b–c = Posidonius frr. 124, 125. On this view the moon is partially translucent rather than opaque (being composed of fire and air, Aetius 2.25.5 = Posidonius fr. 122). This theory is problematic, as Plutarch recognizes, but it does not present a radical alternative to heliophotism. See comments in Edelstein and Kidd 1972–1988, 2:473–80; Préaux 1973: 188–94. The moon is spherical for the Stoics including Posidonius: Aetius 2.27.1 = Posidonius fr. 122. Cleomedes holds that the moon does not reflect solar light but rather is stimulated by it to give off its own secondary light (2.4.21–32). Nevertheless, his theory (like that of Berosus, for the most part, which he criticizes, 2.4.1–17) is empirically indistinguishable from heliophotism, and he explains eclipses, both solar and lunar, by antiphraxis (2.4.127–37; 2.6). Writing in the Stoic tradition, he probably composed his treatise before AD 200, as scholars now believe, rather than in the fourth century AD as used to be thought (see Bowen and Todd 2008). Cleomedes attributes heliophotism to the earliest natural philosophers, 2.5.81–82.

49. Pappus *Mathematical Collection* 6.555–56.

	Solar Eclipses	Lunar Eclipses
Heraclitus	O	O
Anaxagoras	(A)	A
Empedocles	A	(A)
Philolaus	A?	A
Antiphon	?	
Aristotle	(A)	A
Berosus		O
Chrysippus, Stoics	A	
Mathematical astronomers	A	(A)
Aristarchus	A	(A)

From Anaxagoras on, everyone accepts antiphraxis except Berosus.

6.6 PLATO'S HEAVENLY SPHERE

One major philosopher who seems to stay aloof from the new astronomy is Plato. He knows of Anaxagoras' theory of lunar light, which he discusses in the *Cratylus* (409a–b), but only to hint that Anaxagoras has been anticipated in his theory by the ancients who coined the moon's name. Plato is deeply worried about natural philosophy in the Ionian tradition, for it tends to exalt chance and nature in a way that leads to atheism (*Laws* 10.888e–889c et passim). Plato does not directly borrow any astronomy from Parmenides or Anaxagoras.

Yet there are features of his cosmology that suggest indirect influence. In his *Timaeus* he depicts the created cosmos as having a perfect spherical body (33b–c). The Demiurge makes a circle of the Same and a circle of the Different as orbits around the earth that intersect and ultimately account for the daily and annual motions of the sun and some motions of the planets (36b–c). Plato identifies five planets besides the sun and moon, and names the Morning Star and Mercury (38d).[50] The sun illuminates the heavens and seems to be responsible for night and day (39b–c). The Demiurge fashions a "heavenly race of gods," namely the stars, mostly from fire, and makes their form "well rounded" (*eukyklos*), using a word dear to Parmenides.[51] Plato stresses importance of immaterial souls or soul-stuff in the motion of the heavenly bodies.[52] He consciously seems to eschew physical explanations of the heavenly bodies and seems to aim at a more geometrical and teleological account of the heavenly than what he finds in the natural philosophers.

Yet for all his change of emphasis, Plato is *au courant* of the five planets, the sun as the main illuminator of the cosmos, and the sphericity of the heavenly bodies, interpretations that grow out of the researches of the natural philosophers.[53] What is perhaps most significant for the future is the notion of a heavenly sphere. This picture will become standard in Aristotle, who will pass down the conception to later antiquity and the Middle Ages. Plato cites no antecedents for this picture. But the first vision of a heavenly sphere seems to arise in Parmenides' cosmology. Parmenides is the originator of the "centrifocal" universe, as Furley dubbed it,[54] a spherical universe in which rotational symmetry prevails and motions are measured relative to the sphere's center. Anaxagoras does not formally represent his heavens as a sphere, but he does recognize and provide empirical evidence for orbits that circle under the earth, as we have seen (above, section 6.4). This move in effect puts the earth at the center of multiple orbits, which

---

50. Plato notes that most of the planets have not been named (39c). Does that mean the other three planets?
51. B8.43; cf. B1.29, with the reading of Simplicius.
52. *Timaeus* 35a–36d, 38b–e, with Vlastos 1975, ch. 2.
53. The five planets, as we have pointed out, had long been known by Babylonian astronomers. They appear with Greek translations of their Babylonian names in the pseudo-Platonic *Epinomis* (perhaps by Plato's student Philip of Opus), 986e–987d, where a Syrian or Egyptian connection is suggested.
54. Furley 1987:41, 53–54.

taken together potentially make up a heavenly sphere. Later in the fifth century Philolaus finds rotational symmetry in his hestiocentric universe with the "lowest part" of the cosmos corresponding to the "highest part" (B17): "he was the first philosopher-scientist who explored the role of the spherical structure of the universe with the symmetrical notions of up and down" (Maniatis 2009:408).

The heavenly sphere will become the standard picture in Aristotle's *On the Heavens*, with the heavenly bodies being composed of and surrounded by the fifth element, and being changeless except for their circular motion. Aristotle and his contemporaries can now envisage the heavens as a series of concentric spheres and try to describe them in solid geometry. Moreover, Eudoxus developed a theory of spheres with poles oriented differently and moving at different speeds to create a model of the heavens. Callippus and Aristotle went on to make improvements to his model (Aristotle *Metaphysics* XII.8). But they all saw the heavenly bodies as moving with uniform motion along a circle. The results of the peculiar motions of different spheres produced the appearance of complex motions in the planets.

The result of this development was to make astronomy an exercise in kinematics based on spherical geometry. All subsequent ancient astronomy was based on the model of a heavenly sphere. The essential first step was in the process was to allow the bodies to orbit around the earth without barriers. Anaxagoras took the first step in that direction.

6.7 ARISTOTLE'S PARADIGM

It is instructive to see how Aristotle responds to antiphraxis.[55] He does not tell us who originated the theory. (In general he seems to hold that philosophers, led on by the truth itself, inevitably discover,

---

55. For a study of Aristotle's treatment of eclipses, see Goldin 1996, which focuses on epistemological rather than empirical features of the explanation.

or more correctly rediscover, the truths of nature.)⁵⁶ He does, however, stress that the understanding of eclipses is an example of correct scientific reasoning. At the beginning of *Posterior Analytics* II, where his subject is causal explanation, he wishes to show how explanations can be put in the form of a syllogism, with a middle term linking the major and minor terms:

> What is an eclipse? Privation of light from the moon by the earth's screening (*antiphraxis*). Why is there an eclipse? or Why is the moon eclipsed? Because the light leaves it when the earth screens it [*antiphrattousēs tēs gēs*]. (2.90a15–18, trans. Barnes)

Later Aristotle uses the same example to meditate on the question of how we can go from a partial or incidental understanding to a scientific knowledge of an event.[57] We may, for instance, recognize that the full moon is not casting a shadow, but this is not an explanation of a lunar eclipse, only a concomitant. When we realize why it is not casting a shadow, or why it is that the moon is darkened, we have scientific understanding. The middle term that links the moon (the subject of the conclusion) with its property of being eclipsed (its predicate) is the middle term, and this is the interposition of the earth (*antiphraxis gēs*).

There are some gaps in this story. Aristotle does not clearly tell us how we know when we have hit on the right explanation rather than just another concomitant feature of the situation. He does say that if we were on the moon during a lunar eclipse we would perceive the earth blocking the sun's light; but this would not by itself entail a general understanding of eclipses.[58] (In fact an observer on the moon at

---

56. Aristotle *Metaphysics* I.984a18–19, b8–11; Aristotle *Met.* XII.1074b10–13; *Politics* VII.1329b25–30.
57. 8.93a30–b7.
58. I.31.87b39–88a2.

the time of a lunar eclipse would see a solar eclipse.) But for our purposes the important thing is that he thinks there is one and only one correct account of lunar (and solar) eclipses, which is given by antiphraxis.[59] And this theory is so secure that it can act as an exemplar or, to use the word Thomas Kuhn has made fashionable in our own time, a *paradigm* of successful problem-solving.[60] In other words, the antiphraxis theory is real science, a piece of irrefutable reasoning that can be reduced to a simple syllogism: The moon's light (from the sun) is blocked by the earth; the blocking of the sun's light by the earth is an eclipse; therefore, the moon is eclipsed.[61]

It would be nice to know the full range of evidence that Aristotle thought contributed to the correct syllogism embodying the causal role of antiphraxis. This would include the evidence that the moon is a dark body illuminated by the sun (heliophotism), a fact that was vividly confirmed to Aristotle by the occultation of Mars he observed at the time of a half moon. He observed the planet being obscured by the dark half of the moon and later emerging from behind the bright half.[62] Here obviously the moon was acting as a dark body, half illuminated. Aristotle understood how the phases of the moon resulted from shadows on a circular body.[63] He also noted that only a spherical body could produce the crescent-shaped appearance of a solar eclipse.[64] Eclipses occurred as predicted, with solar eclipses occurring at the new moon, lunar eclipses at the time of the full moon. He understood how the circular shadow cast on the moon during a lunar

---

59. He takes the syllogism, properly laid out, to explain the cause, and to form the basis of a real definition of eclipses: *Post. An.* 98b16–24; *Met.* 1044b9–15.
60. Kuhn 1996:10–11 et passim. Here I coopt the term and the general historical phenomenology of Kuhn, without the radical (and, I think, unnecessary) antirealist conclusions he draws.
61. For Aristotle the knowledge conferred is timeless demonstrative knowledge, not merely knowledge relative to a historically restricted paradigm.
62. *On the Heavens* 292a3–6.
63. *On theHeavens* 291b18–21.
64. *On theHeavens* 291b21–22.

eclipse proved the sphericity of the earth—a view not attested among the Presocratics, but a further application of Anaxagoras' theory "according to sense experience."[65] Thus there were multiple ways to confirm the correctness of heliophotism and antiphraxis, so that one might empirically justify the logical connection of antiphraxis between the subject "moon" and the predicate "eclipsed."

There is one incongruous feature in Aristotle's account of eclipses: it fits poorly with his own theory of how the heavens impart light and heat to the Earth. According to Aristotle, the motion of (some) heavenly bodies causes the upper atmosphere (below the moon) to become heated or inflamed, and that disturbance of the atmosphere heats up the Earth.[66] But from the moon outward to the outermost heaven everything is composed of the fifth element, which does not undergo any changes other than circular motion.[67] How then can one heavenly body generate light and another reflect that light? If the motion of the sun by itself causes light and heat, how can the moon screen its effect in a solar eclipse? And why is the moon not a source of heat as well as (sometimes) of light, given that it is closer to the upper atmosphere? These questions are difficult to answer, and they seem to be impossible to reconcile with a theory that makes the sun the sole source of light in the lower part of the heavens. As Andrew Gregory notes, "The ignition theory in *De caelo* is simply incompatible with the implicit assumption [Aristotle] often makes that the Moon shines by reflected light."[68] This creates an awkward situation for Aristotle, whose cosmology seems to be in conflict with his astronomy. But for our purposes, it suffices to note that Aristotle takes as a paradigm of knowledge a theory of the heavens which does not fit well with his own cosmology. It appears that the evidence for

---

65. DC *Posterior Analytics* 297b23–30.
66. *On the Heavens* II.7; *Meteorology* I.3.
67. *On the Heavens* I.2–3.
68. Gregory 2000a:249; see the accompanying analysis.

eclipses was so overwhelming that it never occurred to him to question it, even when it could not be reconciled with his own physical theories.

As a footnote to Aristotle, we see an application of heliophotism in the pseudo-Aristotelian *Problems* (15.7):

> Why, though the moon is spherical, do we see it straight when it is half-full? Is it because of vision and the circumference of the circles which the sun makes when it falls upon the moon are in the same plane? . . . When [the circle illuminated by the sun's light] is opposite to us the whole is visible and the moon appears to be full; but when it changes owing to the altered position of the sun, its circumference becomes on a plane with our sight and so it appears straight. . . . (911b35–912a7, new Oxford trans.)

The author of this Hellenistic work goes on to describe correctly how the moon's phases result from the angle of the sun's light falling on the moon as seen from the perspective of the observer. We see heliophotism becoming part of textbook science in the Aristotelian tradition.[69]

6.8 A SCIENTIFIC CONSENSUS

However Anaxagoras arrived at his theory of the heavens precisely, we have already seen that he had the insight of Parmenides to guide him. He grasped the implications of the theory for cosmology and astronomy, and in particular saw a potential application for explaining eclipses. By the time of Aristotle his theory had proved so successful that it led to new insights, such as a proof for the sphericity of the earth (which Anaxagoras himself had denied). It had provided

---

69. Compare the hypotheses of Aristarchus in the epigraph to this chapter.

what Imre Lakatos would call a research program with a consistently progressive theoretical problemshift.[70] It had emerged as what Aristotle could recognize as a paradigm of good science.

We have followed Anaxagoras' theory and its competition through the doxographic reports that are designed not to illuminate the history of philosophy or of ideas, but to show their dialectical interrelations for the purpose of further argument. Although these sources tend to obscure the historical significance of theories, we have found that when corrected they vindicate Anaxagoras' theory in a surprising way. They show that after he put forth his hypotheses, debate on the topic virtually ceased. Subsequent theories were restatements of his theory. There emerged one of the most elusive of all phenomena in Greek doxography: a consensus.

As far as we can see, Anaxagoras' theory caught on almost immediately: embraced by Empedocles, Philolaus, at least in part by Diogenes, Antiphon, and Democritus, it arrived at Aristotle as an already well-confirmed theory. It was picked up by the Stoics, from Zeno down to and including Posidonius, and also and more important by scientific astronomers including Aristarchus, Hipparchus, and subsequently Ptolemy, and from there by anyone who knew Ptolemy. It is striking that heliophotism and antiphraxis survived radical changes in perspective: from geocentric theories to a hestiocentric theory (Philolaus) to heliocentric theories (Aristarchus, Copernicus[71]); from theories of freely moving heavenly bodies (many of the Presocratics) to nested spheres (Eudoxus, Aristotle, Callippus) to epicyclic theories (Apollonius of Perge, Hipparchus, Ptolemy) to theories of elliptical orbits (Kepler).[72] We have found only one holdout to the view, the Babylonian Berosus, who at least preserved many features

---

70. Lakatos 1970:134.
71. E.g. *De revolutionibus* 4.18, 4.31.
72. See Johannes Kepler, *Epitome of Copernican Astronomy*, Book 1, Part 1, ch. 4, pp. 480–81.

of heliophotism, while rendering the correlations accidental.[73] The Epicureans for their part withheld judgment as a matter of principle on this as other theories dealing with remote phenomena. While almost all theories of the Presocratics remained controversial and generated alternative hypotheses, this set of views put an end to new speculations on the subject. It became the norm for the rest of history—from the early fifth century BC to the present day. While no doubt many subsequent theorists accepted the view on authority, many other astronomers subjected it to the most rigorous scrutiny, at least from Aristotle's time until Galileo's, and no one ever found another theory which could match it in accounting for the phenomena.

It is true that if one reads only doxographical and philosophical literature, one can get a different sense. Late in antiquity, for instance, Augustine says that it is impossible or difficult to know whether the moon shines by reflected light or whether half of the moon has its own light (Berosus' theory).[74] But of course the two theories are empirically equivalent in their predictions. Astronomers typically saw the latter as an *ad hoc* and scientifically empty substitution of the first. But to non-astronomers the points of dispute could seem baffling and undecidable.

The dominance of antiphraxis theory among astronomers in ancient times is evident already in Geminus' *Introduction to Astronomy* of c. AD 50:

> Eclipses of the sun arise from the moon's occultation (*epiprosthēsis*) of the sun. For as the sun travels higher though the sky with the moon below it, whenever the sun and the moon come to be in the same position, the moon as it passes under the sun blocks (*antiphrattei*) the rays of the sun traveling toward us.

---

73. For some criticisms, see Cleomedes 2.4.10–17, and n. 40 above.
74. *On Psalms* 10.3; cf. Isidore of Seville *On Nature* 18.

That is why these phenomena should properly be called not "eclipses" but "occultations": for the sun will never give out (*ekleipsei*) even in part, but it becomes invisible to us because of its occultation by the moon. ... The main evidence that the sun is eclipsed by an interposition of the moon is (1) the fact that it does not happen on any other day than the 30th [day of the lunar month], when the moon is in conjunction with the sun, and (2) the fact that the magnitude of an eclipse is proportionate to the viewer's proximity to the path of the eclipse. (10.1–2, 6)

Eclipses of the moon arise from the earth's entering into the shadow of the moon. For just as other bodies illuminated by the sun cast a shadow, so also the earth, which is illuminated by the sun, casts a shadow. ... In each case as part of [the moon] enters the shadow of the earth it loses its light from the sun because of the interposition of the earth. For at that time the sun, the earth, the earth's shadow, and the moon are aligned along a straight line. The reason eclipses of the moon do not happen on any other day than the full moon is that only then are the sun and moon in opposition. (11.1, 3–4)

This textbook, offering "a well-organized and more or less complete introduction to astronomy,"[75] embodies the science of the time, and explicates eclipses as resulting from interpositions; the nature of eclipses is a settled matter of fact without doubt or controversy. The causes, geometry, and timing of eclipses are part of the rudiments of astronomy any student must learn to understand the relevant phenomena. Geminus goes on to point out, correctly, that lunar eclipses are equally visible to everyone who can see the moon at the time (11.5), unlike solar eclipses, which are visible only to those who are near the line connecting the earth, the moon, and the sun (10.3–5).

75. Evans 1998:91.

The days of the month on which eclipses occur provide evidence for the truth of the theory. The fact that solar eclipses can be seen only by those in or near its path confirms the theory. These are points Anaxagoras grasped in the early fifth century BC. In the first century AD they have become facts learned in an introductory textbook.

We may compare a contemporary exposition of eclipses:

> A solar eclipse occurs when the Moon passes in front of the Sun as seen from the Earth.... A total solar eclipse is in fact an occultation while an annular solar eclipse is a transit.... Lunar eclipses occur when the Moon passes through the Earth's shadow. Since this occurs only when the Moon is on the far side of the Earth from the Sun, lunar eclipses only occur when there is a full moon. Unlike a solar eclipse, an eclipse of the Moon can be observed from nearly an entire hemisphere.[76]

While our understanding of the physics of celestial motions, the composition of the heavenly bodies, the shape of orbits, and the velocities of the bodies has increased greatly since antiquity, the basic facts of eclipses have remained unaltered for almost 2,500 years, and have been common knowledge for astronomers for at least 2,000.[77]

---

76. Wikipedia, s.v. eclipse, http://en.wikipedia.org/wiki/Eclipse, July 11, 2011.
77. I assume that Geminus did not write the first primer of astronomy, but only the best known one.

Chapter 7

The Geometry of the Heavens

> SALVIATI: The moon certainly agrees with the earth in its shape, which is indubitably spherical. This follows necessarily from its disc being seen perfectly circular, and from the manner of its receiving light from the sun. For if its surface were flat, it would all become covered with light at once....
>
> In the second place, it is itself dark and opaque like the earth, by which opacity it is fitted to receive and reflect the light of the sun; for if it were not so, it could not do this.
>
> Third, I hold its material to be very dense and solid, no less than the earth's, of which a sufficiently clear proof to me is the unevenness of the major parts of its surface, evidenced by the many prominences and cavities revealed by the aid of the telescope.
>
> —Galileo *Dialogue Concerning the Two Chief World Systems*, First Day (trans. Drake)[1]

Thus far I have presented an argument for the development of early Greek astronomy. This argument has been laid out in part in historical order. But I have provided an argument rather than a historical

---

1. Drake 1967:62–63. I take it that Galileo recapitulates Anaxagoras' (and probably Parmenides') reasoning. His major new contribution is the use of the telescope to confirm the impression of mountains and valleys (his recognition that the earth is spherical follows Parmenides but not Anaxagoras).

narrative of what happened. The point has been to reconstruct the steps of early Greek astronomy so as to identify key turning points in explanations. Now it is time to go back and tell the history of astronomy, briefly, making use of the reasoned reconstruction already defended.

7.1 THE STORY OF EARLY GREEK ASTRONOMY

The early philosophers inherited from their culture a mythological picture of the world in which a disk-shaped earth was surrounded by a heaven above it and an underworld below it. The gods dwelt in the heavenly regions and certain titans and underworld deities in Tartarus below. The heavenly bodies, or at least the more prominent among them, were identified with gods. Helios the sun drove his shining chariot across the sky every day. Selene the moon governed the night time. Day and Night took turns occupying the upper realm on alternating shifts, and retired to the same house to rest during their off hours. Heaven, Earth, Tartarus, and various cosmological features were themselves deities with genealogies and family connections. The gods controlled the earth and sky and could create prodigious events such as earthquakes (Poseidon, god of the sea), lightning and thunder (Zeus, god of the sky), and winds (Boreas and other wind gods). When meteorological, geological, or biological disasters happened, it was a sign of the displeasure of the gods, and a seer or prophet could be consulted to tell mortals how to placate the offended deity.

Convinced that the world was a series of natural events occurring in the natural world governed by natural laws or principles, the first philosophers sought to explain heavenly events on the basis of natural processes. At first they had no clue as to what the real nature and causes of heavenly events were, but they had faith that some sort of natural explanation could account for them. Thales was the first in the line of natural philosophers, but we do not have detailed information about his cosmological theories. He seems to have viewed the earth as like a raft floating on a primordial sea. He may have brought back the concept of a 365-day year from Egypt, and from the Phoenicians a knowledge of Ursa Minor as a constellation at the north pole of the heaven for navigation purposes.

Anaximander developed a complex model of the cosmos with the earth as a disk staying in place by its equilibrium. The heavenly bodies were rings of opaque air filled with fire; the inner fire was visible at an opening where its light shone out. For his successor Anaximenes the heavens were like a felt cap rotating around the head. The heavenly bodies did not set under the earth but were hidden by high mountains to the north. The fixed stars were attached to the heaven, but the sun and moon were thin leaf-like bodies floating freely, driven around by winds or a whirlwind.

Xenophanes posited a flat earth that stretched infinitely on horizontally and extended infinitely down vertically. The heavenly bodies were cloud-like formations. The sun and moon traveled straight on to the west and were replaced by a different sun and moon from the east. There were also different suns for regions far to the north and south. Solar eclipses were caused by the sun's running out of vapor to fuel its fires. Heraclitus understood the sun and moon to be bowl-like structures full of fire. The phases of the moon and eclipses are caused by rotation of the bowls to present the dark side to the earth. The sun is new every day, at least because the fire in it is newly ignited. All of these early theories present the heavenly bodies as relatively light structures, sometimes temporary in existence. The earth is flat. Heavenly bodies do not travel under the earth but around it. Heavenly phenomena are not in principle different from meteorological events, fueled by evaporation from the earth, pushed around by winds, staying aloft because of their light weight and loose structure.

In his cosmology, whatever its purpose, Parmenides challenged the conventional thinking on astronomy. He arrived at three striking insights: the moon gets its light from the sun, the earth is spherical, and the morning star is identical to the evening star. The first insight leads to the second: the shadows on the face of the moon show that it is spherical in shape. By an inductive generalization, all heavenly bodies are spherical, perhaps including the earth. Viewing the moon as a solid, permanent body may have encouraged Parmenides to look at the morning star and evening star as permanent structures also; since only one appeared at any given time, they were likely to be different manifestations of the same permanent body.

Inspired by these insights, Anaxagoras grasped the potential of heliophotism to explain much more than the moon. He saw that the theory entailed that the moon was spherical; that the moon orbited below the sun; that it was solid and opaque; that some heavenly bodies traveled under the earth in their orbits; and that heavenly bodies were permanent features of the heavens possessing weight. He also saw that the hypothesis of heliophotism could potentially account for eclipses: a large heavenly body blocking the sun's light would cause the moon to be darkened, and the earth was in a position to block the sun's light at the time of the full moon. The moon was the same apparent size as the sun and, as a dark body, lurked in the vicinity of the sun at the time of the new moon. While he was in Athens in 478 BC Anaxagoras observed an annular eclipse of the sun. He realized that, as entailed by theory, the event took place at the time of the new moon. In accordance with heliophotism, the screening of the sun by the moon must have cast a shadow on the earth, where the complete eclipse was visible. He could even determine the extent of the umbra, by interviewing sailors and merchants at the Piraeus soon after the eclipse and by corresponding with other star-gazers. By doing this he determined that the umbra covered the Peloponnesus but did not extend beyond it. Given that the size of the moon's shadow approximated the diameter of the moon, he inferred that the moon was the size of the Peloponnesus and the sun larger (since its outline was still visible during the annular eclipse).

Anaxagoras inferred that the heavenly bodies were earthy or stony, as the moon's reflective and opaque surface indicated. In order to account for lunar eclipses that occurred at the rising or setting of the moon, when the flat earth could not account for an eclipse, he posited the existence of other dark bodies, asteroids, below the moon. Only a very powerful circular motion could account for heavy bodies in orbit, so Anaxagoras posited a strong vortex motion, able to pick up boulders from the earth's surface at the earth's periphery. Any disturbance to the bodies in orbit could send a star plummeting back to the earth's surface. Anaxagoras published his views, which won him immediate fame. He was the toast of Athens, where Aeschylus repeated his theory of the Nile floods to fascinated theater-goers.

In or around 466 a meteor fell to earth near Aegospotami, leaving a trail of flames. It occurred at the time a comet, possibly Halley's Comet, was seen in the sky. The meteorite was found, a huge boulder with a burnt color. The event was immediately reported as a confirmation of the young philosopher's theory. A large rock had fallen from the sky, as predicted; it may have been caused by the friction of two asteroids in collision, which had manifested themselves as a comet. The heavenly disturbance knocked another asteroid out of orbit and caused it to fall to earth. Or perhaps Anaxagoras' theory had only portrayed comets as a confluence of the light of planets not in contact. In any case, there was now evidence on the ground of stony bodies in the sky. Anaxagoras' theory had a received a startling confirmation.

As news of the successes of the new theory spread, it was adopted by contemporary philosophers, first Empedocles, then Diogenes. Anaxagoras' general astronomy was followed by Democritus. Even innovative theorists like Philolaus, who made the earth a planet rather than the central body of the heavens, retained the insights of

Anaxagoras and Parmenides. The vortex theory became the standard account of heavenly dynamics and the earthy or stony composition of heavenly bodies the standard astrophysics. The Meteorological Model of the pre-Parmenidean philosophers was rendered obsolete at a stroke and replaced by the Lithic Model. The heavenly bodies were three-dimensional, earthy or stony, permanent bodies, some of which traveled under the earth. The sun was the cause of much or all light in the cosmos, including daylight. The moon might have some dim secondary light of its own, but its phases were caused by reflections of sunlight. The existence of eclipses was explainable on the basis of celestial geometry without recourse to either supernatural interventions or special physical circumstances. As Aristotle recognized over a century later, Anaxagoras' theory was the kind of explanation that generates its own empirical verification and entails far-reaching consequences for a field of study.

In the later fifth-century, however, two movements interrupted the advance of astronomy. One was the sophistic movement, with its emphasis on practical applications of knowledge, especially in the political sphere. Some sophists (such as Hippias and Antiphon)[2] participated in cosmological speculation. Others, however (for instance Protagoras),[3] regarded cosmology and astronomy as worthless distractions from the useful and productive areas of knowledge. On the positive side, sophists provided a kind of humanistic emphasis that was absent from earlier education. On the negative side, they might dismiss natural philosophy as unverifiable speculations about things far beyond human ken. The irony of this last accusation should now be evident: just at the time when philosophical speculation was developing scientific tools to ground its scientific claims, it was charged with having none.

2. Hippias: Philostratus *Life of the Sophists* 1.11.1 = 86A2; Plato *Hippias Major* 285b–c = A11. Antiphon: Aetius 2.20.15 = 87B26; Aetius 2.28.4 = B27; Aetius 2.29.3 = B28.
3. Plato *Protagoras* 318d–319a = 80A5.

THE GEOMETRY OF THE HEAVENS

The other fifth-century movement was a conservative backlash against the new thinking—provoked at least in part by agnostic and subversive claims made by some sophists. Threats of prosecution for impiety apparently drove Anaxagoras out of Athens, and brought all philosophers under suspicion of being dangerous radicals. Perhaps just as Anaxagoras' scientific theories were coming to be widely accepted, he suffered a kind of political and religious censure. In this intellectual environment, Socrates challenged the wisdom of contemporary wise men, both sophists and citizens of Athens, and sought for a new basis for moral discourse. He had himself been a student of the natural philosopher Archelaus,[4] but he became part of a humanistic reaction to natural philosophy, and eventually a victim of the anti-intellectual backlash in Athens. While Xenophon portrays him as hostile to natural philosophy, Plato portrays him as simply indifferent.[5]

After his death, Socrates' followers continued his crusade as each understood it. Although it seemed as if natural philosophy might be banished from the new philosophy, Plato offered a way of rehabilitating the study: it could be subordinated to questions of how things were arranged for the best (*Phaedo* 96a–100a).[6] Plato cited Anaxagoras as the inspiration for this program: if Mind organized the world, the features of the cosmos must be disposed in the best way possible. He went on to register his disappointment that Anaxagoras did not exploit his own great insight. Yet the insight provided a starting point for scientific inquiry from within a humanistic or idealistic framework. If, that is, natural philosophy could be reconceptualized so as to address questions of value, it would no longer be a threat to morality and goodness. This program was carried out in a tentative way

---

4. Diogenes Laertius 2.23 with Graham 2008.
5. Xenophon *Memorabilia* 1.1.11–16; Plato *Apology* 19b–d. I take the *Phaedo* to reflect Plato's views rather than Socrates'.
6. See Graham 1991.

in the *Timaeus* and in a more confident and systematic way by Plato's student Aristotle. Plato's own astronomical observations were limited, but he did seem to accept the concept of a spherical earth, surrounded by heavenly bodies that move according to mathematical regularities.[7]

Aristotle followed a mathematical model of the heavens first worked out by his contemporary Eudoxus in the Academy.[8] Aristotle, as we have seen, accepted heliophotism and antiphraxis as fundamental and indeed paradigmatic discoveries in astronomy. For philosophical reasons he rejected earthy bodies in the heavens. According to him all bodies from the moon outward traveled in the realm of the fifth element, called aether by the ancients, where there was no change except for circular motion.[9] Thus all the heavenly bodies were everlasting and constant in their movements. Accordingly, meteors could not be heavenly bodies, but meteorological events from the sublunary world. Below the moon the four elements (earth, water, air, fire, the basic realities identified by Empedocles) changed into one another, making up the compound bodies of our world.[10] The fiery vapors of the upper atmosphere (below the moon) could change into earthy material, producing a meteor, which because of the behavior of earthy material would fall to earth.[11] Thus there were meteors, as observers had found out in the time of Anaxagoras, but they arose from meteorological processes, not from astronomical events. Similarly comets arose in the sublunary atmosphere.[12] The composition and behavior of heavenly bodies were utterly different from

---

7. Plato *Phaedo* 110b–c; *Republic* 10.616b–617b; *Timaeus* 34a–b, 36b–e, 38c–e. See Cornford 1937: 80–89, 106–112; Dicks 1970:122–30; Vlastos 1975, ch. 2; Gregory 2000b, ch. 4.
8. Aristotle *Metaphysics* 1073b17–32; Simplicius *On the Heavens* 488.18–24, 493.4 ff. with Heath 1913:190–211; Kuhn 1957:55–59.
9. *On the Heavens* I.2–3.
10. *On Coming To Be and Passing Away* II.7–8.
11. *Meteorology* I.4–5.
12. Aristotle *Meteorology* I.1.338a25–b24; I.6–7.

those of sublunary bodies, and astronomy was definitively distinguished from earth science.

In retrospect Aristotle took a step back from the naturalistic view fostered by Anaxagoras: instead of a single type of matter and a single set of law-like regularities, Aristotle posited a realm of aetherial bodies following special laws of motion. Yet Aristotle's picture had one significant advantage: the heavenly sphere made up of circular orbits was difficult or impossible to account for on the basis of the physics of the Lithic Model. A vortex is essentially a cylinder or inverted cone that should drive bodies caught in it around the earth in one plane, or at least in parallel planes; but the heavenly bodies move around the earth in intersecting planes. The "centrifocal" universe needs a physics that can allow bodies to move about the earth in three dimensions. The theory of nested spheres allowed for relatively free circular motions around the earth and in that important respect was superior to the vortex model.

The mathematical models of Aristotle's time, using nested spheres, were developed by Eudoxus, Callippus, and Aristotle.[13] Around 280 BC Aristarchus proposed a heliocentric model, which failed to persuade his contemporaries. Subsequently Apollonius of Perge (c. 200) developed the methods of eccentrics and epicycles to account for variations in the movement of heavenly bodies. Hipparchus (c. 135) drew on Babylonian data—which he seems to have searched out and introduced to the Greek tradition—to provide accurate models of heavenly motions which could predict eclipses.[14] Claudius Ptolemy (c. AD 150) produced an elegant theoretical account of heavenly motions, incorporating ancient data and earlier theoretical models.[15] All through this period heliophotism and

---

13. Aristotle *Metaphysics* XII.8 with Heath 1913:190–224.
14. See Toomer 1988.
15. See Kuhn 1957:59–73.

antiphraxis were assumed and described geometrically. Aristotle's cosmology remained the dominant model of physics, but it could not account for eccentric or epicyclic motion of bodies, so the mathematical models of astronomy remained unreconciled with the physics of the cosmology.

At this point some scholars will complain that I have left out of my story the real advances made empirical astronomy: Meton made precise observations of the solstice of 432 BC and developed an accurate lunisolar calendar of nineteen years consisting of 235 lunar months, while his contemporary Euctemon distinguished the fact that the astronomical seasons are not all equal.[16] Their work is supposed to mark "the real beginning of mathematical astronomy in Greece...."[17] Unfortunately for this view, the alleged observational and mathematical advances can all be explained more plausibly as occasional borrowings from Babylonian sources. The nineteen-year cycle corresponds closely with the so-called Uruk scheme used by Babylonian astronomers and is likely borrowed from them. The day of the summer solstice for each year is defined by the Babylonian scheme and need not be determined observationally. Meton and Euctemon probably developed *parapēgmata* or almanacs, and Euctemon may have simply used his to determine the seasons in an a priori way.[18] In fact, Theophrastus says that Meton learned the nineteen-year cycle from a foreigner named Phaeinos.[19] This reconstruction does not support the attribution of a high level of observation and mathematical analysis to Meton and Euctemon. Even the borrowings do not suggest a sophisticated knowledge of Babylonian astronomy,

---

16. Mentioned briefly above, ch. 1, introduction, and section 1.1.
17. Dicks 1970:89.
18. Bowen and Goldstein 1988. Lehoux 2007:89 is more cautious, but offers no evidence against this explanation. Bowen and Goldstein show that even lunisolar cycles that are not borrowed can be calculated simply on the basis of the difference between a solar year and twelve lunar months, without precise observations.
19. Bowen and Goldstein 1988:80; Theophrastus fr. 6.4 Wimmer.

which was probably achieved only by Hipparchus.[20] On this reconstruction, the alleged competition for the title of the founder of scientific astronomy among the Greeks is much weaker than has been thought.

The case for observational astronomy in Greece gets even worse. Eudoxus is credited not only with a sophisticated mathematical model of the heavens, but with the first precise observations of stars and constellations. His writings are lost, but his data informed Aratus' poetic star chart, *The Phaenomena* (third century BC). Because of the precession or circling of the earth's axis, the location of constellations in the sky gradually changes over time. Studying Aratus' descriptions of the constellations, archaeoastronomer Bradley E. Schaefer (2006) finds that the observations he reports could only have been made around 1130 BC at 36° N. latitude—about the time and place the MUL.APIN tablets originated. In other words, Eudoxus must have borrowed his "observations" wholesale from a Near Eastern source without even checking their accuracy! If this is right, then the state of observational astronomy in Greece was dismal through the fourth century BC.

7.2 SCIENTIFIC PROGRESS

It appears, then, that the celebrated observational astronomy of Meton and Euctemon in the last third of the fifth century BC amounts to low-level borrowings from the East, prefiguring more unacknowledged borrowings in the following century; whereas the maligned theoretical astronomy of Parmenides and Anaxagoras in the first third of the fifth century constitutes the real foundation of Greek scientific astronomy. A continuous thread leads from Parmenides and

---

20. See above, n. 14.

Anaxagoras to the present day. For almost 2,500 years no other theory of lunar light and eclipses has been able to match the one proposed by Anaxagoras on the basis of an earlier theoretical insight by his predecessor Parmenides and a few naked-eye observations. It is still found in astronomical handbooks and always will be. In 1969 astronauts from planet earth walked on the moon for the first time and found it to be a solid, opaque, earthy body—with mountains and valleys—of roughly spherical shape, illuminated by the sun and orbiting around the earth. Of course, they were not really trying to confirm these facts, which were already foregone conclusions, but more difficult questions of the age and composition of moon rocks and so on. But this just goes to show that the claims made by Anaxagoras had been settled long before the Apollo missions landed, and settled in favor of the Presocratic theorist. In fact, studies of lunar minerals showed that the moon was earthy in a very literal sense: the minerals contained the signature of earthy composition, showing that the material in the moon had been ejected from the earth probably in a massive collision with a planet-sized body early in its history.[21]

Are scientific theories cumulative? This is a big question, but the answer that emerges from this small study is that at least sometimes they are. There is one astronomical theory that has a track record of success for two and a half millennia. Other questions that Anaxagoras speculated about have been replaced by other theories, sometimes theories unimaginable to the Presocratics. For instance, it was only in the twentieth century that the composition of the sun and its source of radiation were determined: the sun is a ball of mostly hydrogen that undergoes nuclear fusion in its core. It is not a molten mass of rock heated by friction or rays of aether. Without an advanced knowledge of gravitational forces and nuclear physics the problem of

---

21. It is now thought that this planet, known as Theia, traveled in roughly the same orbit as the earth—a kind of counter-earth like that Philolaus envisaged. See http://en.wikipedia.org/wiki/Giant_impact_hypothesis.

solar light was insoluble. Yet that the sun was the source of light for the earth and moon was already known long since.

Was what Anaxagoras did astronomy? Why not? The biggest objection seems to be that Anaxagoras and his generation did not propose mathematical models of planetary motion (ch. 1.1 above). But while it is true that mathematical models of planetary motion are instances of astronomical theory, not all astronomical theory focuses on mathematical models of planetary motion; being a mathematical model of planetary motion is a sufficient but not necessary condition for being scientific. Thus it is irrelevant, logically irrelevant, that the theory does not instantiate some *other* research program. What Anaxagoras' theory does do is to clarify the relationship of the sun and moon, and, in the process, establish other basic facts about the heavens. The heavenly bodies are permanent, massy bodies, of roughly spherical shape, some of which travel under the earth in their orbits. One body radiates light while another reflects that light; the second is in a lower orbit, the shadows on its surface accounting for its phases and its disappearance from view once a month.

But secondly, the theories of lunar light and eclipses *are* mathematical theories! We can see this for instance in the way that Aristarchus gives heliophotism as a hypothesis for the deduction of truths about the heavens, using geometrical methods.[22] Eclipses are consequences of heliophotism and other assumptions about the heavens, which can be modeled rigorously in solid geometry. The correct parameters were not known to Anaxagoras, nor were they to Aristarchus.[23] But what makes the theory mathematical is not a control of all the sizes and distances—or shapes of the heavenly bodies—but the use of a geometrical method. No one disputes that Aristarchus produced scientific theories. Anaxagoras' geometry was no doubt

---

22. See epigraph to ch. 6.
23. Aristarchus' results are notoriously erroneous. See above ch. 2 n. 52 and Heath 1913, pt. 2, ch. 3.

crude and informal in comparison to Aristarchus', but Anaxagoras did grasp the relevant geometrical relationships, which were elegantly and deductively formulated by his Hellenistic successors. He saw that the phases of the moon are the result of an appearance of the moon as it revolves around the earth with half of its surface illuminated by the sun; he saw that lunar eclipses result from an interposition of the earth between the sun and the moon, and solar eclipses result from an interposition of the moon between the sun and the earth. He saw how everything looked from *outside* the earth—a cosmic geometry, a God's-eye view of the cosmos that made mysterious appearances ordinary consequences of the structure of the world. Anaxagoras took Parmenides' insight and followed its mathematical and physical implications. This is a feat comparable to the more celebrated intellectual leaps of Aristarchus and Copernicus—for which it prepared the way.

If we criticize fifth century astronomy for not offering a mathematical model of planetary motion, we put ourselves in the position of criticizing fifth century astronomy for not being fourth century astronomy. We might as well complain that it does not offer a heliocentric theory of the solar system. Indeed it does not, but we should not expect it to until certain other problems are solved. What it accomplishes is to provide the conditions under which a later generation of astronomers could raise problems like that of how to show that planetary motions are regular. The history of science should point us to an appreciation of what is historically possible at a given stage of development. If the preceding story is historically plausible, fitting with the evidence we have for the fifth century, it indicates that Anaxagoras discovered reliable truths about the heavens that laid the groundwork for the mathematical astronomy of the fourth century. If the astronomy of the fourth century is demonstrably scientific, and if it based on scientific discoveries of the fifth century, then the astronomy of the fifth century is scientific.

One major objection to this argument is that heliophotism and antiphraxis are just low-level observational accounts that do not rise to the level of scientific theories. They come close to being basic facts that are accordingly not part of the high-level explanation expected of an astronomical theory. To this the proper answer is that, as we have seen, in the time that Anaxagoras proposed them, these explanations were risky, implausible, unanticipated hypotheses that made unexpected connections among astronomical phenomena. A priori Anaxagoras' explanatory hypotheses were no more likely than any of their competition. That they constituted both high-level explanations and successful theories, we have Aristotle himself as witness.[24] Even early critics of cosmology show that in the fifth and early fourth centuries astronomical phenomena seemed impossibly remote and their explanations irresponsibly speculative—that is, too advanced for human comprehension.[25] If they now seem to be low-level facts, that is because they have survived centuries of rigorous testing and become an indispensable part of the foundational assumptions of astronomy. In their time they were breath-taking leaps of scientific imagination. And they opened up a way of thinking about the heavens that demanded geometrical models and mathematical computations. That is, after all, what a good scientific theory should do.

7.3 HISTORICAL AND PHILOSOPHICAL SIGNIFICANCE

Francis Bacon long ago anticipated the view of modern historians of philosophy and science when he said,

> [T]hat wisdom which we have derived principally from the Greeks is but like the boyhood of knowledge, and has the

---

24. See above, ch. 6.7.
25. See above, ch. 1.1; Epicurus *To Pythocles* 85–87, *To Herodotus* 79–80.

characteristic property of boys: it can talk, but it cannot generate; for it is fruitful of controversies but barren of works.[26]

Bacon's contemporary Galileo recapitulates ancient astronomical truths without acknowledging any indebtedness to them.[27] The present study has been a story of how, notwithstanding evidence and presumptions to the contrary, Greek wisdom did produce tangible works—a real understanding of astronomical bodies and their relationships. The thesis that emerges from the story is anything but new. It seems to have been held already by Plutarch, and to have been recovered in the later nineteenth century.[28] But the thesis has never won out as the dominant interpretation, and it has often seemed trivial and inconsequential even to those who accepted it. If the present reassessment is correct, Anaxagoras played a key role in putting astronomy on a scientific footing. At least part of the value of telling the story is to show how, in their time and place, the twin theories of heliophotism and antiphraxis were anything but trivial. They allowed astronomers to see the heavens as an orderly place furnished with permanent bodies moving along determinate paths. If the order was not yet mathematical, it was nevertheless regular enough to suggest explanations that could be confirmed empirically. To call the theory of Anaxagoras trivial is to fail to appreciate the status quo ante and the status quo post, to fundamentally misunderstand history. Anaxagoras profoundly changed the understanding of the heavens irreversibly and forever.

This story abounds in ironies. If it is true, two of the most theoretical of philosophers, Parmenides and Anaxagoras, stand at the

---

26. *The Great Instauration*, preface (published 1620).
27. See epigraph above and n. 1. Galileo is arguing against Aristotelian conceptions and focusing on Aristotle's shortcomings.
28. Görgemanns 1970:35–38 points out that Plutarch alone defended the contributions of Anaxagoras against the prevailing systems of his time. For modern contributions, see below, Appendix 1.

fountainhead of Western science. The Presocratics who are reputed to have had their head in the clouds turn out to have produced the first concrete advances in scientific reasoning. And they did it in the one area that was, by all accounts, most remote from human observation and verification. Those theories for which observation was most accessible, what we would today call chemical theories, were some of the most intransigent, waiting until the nineteenth century for their resolution.

It has been popular in recent decades to see science as an all-too-human enterprise in which knowledge is constructed rather than discovered. No doubt there is much to be said for the new approach, and much that is naive in thinking that scientific knowledge is just waiting to be discovered. Yet there is a danger in the new approach of thinking that everything is up for grabs, that almost any interpretation of the world will do equally well. In the present case we see that there were many ingenious hypotheses concerning cosmological and astronomical phenomena. Each could claim some initial plausibility and point to some confirming observations. Nevertheless, only one hypothesis of the moon's light really connected with the progression of the moon's phases and the relative position of the moon. And that one hypothesis also suggested an elaborate model of the heavens, which in turn intimated a hypothesis of solar and lunar eclipses. When this hypothesis was tested against actual eclipses, it accounted for them in most particulars; and with further refinements, such as recognizing that the earth was spherical, it was able to account for eclipses in all particulars.

The record of the reception of this new theory, as well as we can reconstruct it from the rather promiscuous reports we get, shows that it persuaded virtually all contemporaries and all successors. This reception does not seem to have resulted from mere chance, or from some kind of community pressure or scientific politics. For at this time science had virtually no economic or political clout and only dubious intellectual prestige.[29] If Anaxagoras had the advantage of a powerful patron, Pericles, that did not guarantee that his ideas would prevail outside Athens; indeed, it may have made him a target of prosecution by Pericles' enemies in Athens.[30] Science was not organized or institutionalized, and furthermore the claim that philosophy was organized into schools from the Presocratic period on seems to be anachronistic.[31] In a time of intellectual pluralism—we might say anarchy—ingenious theories typically came and went without persuading hardly anyone. If Anaxagoras' astronomical theory prevailed, it did so because it offered genuine advantages in competition with all other explanations. We see that in the realm of physical science all hypotheses are not created equal. Some of them provide powerful explanations of phenomena, and connect information in ways that have not been recognized before. A promising hypothesis makes important connections and suggests further ways of testing the theory.

There are a number of influential theories of scientific development that suggest that genuine scientific progress should be impossible or at least highly unlikely. New paradigms are incommensurable with old, and the accumulation of knowledge is impossible across paradigm shifts. Yet here we find one set of explanations that survived important reorientations in scientific explanation, from geocentric to

---

29. See Lloyd 1996, esp. 213–23, contrasting the social-political context of Greek and Chinese science. The lack of institutional support of Greek science was even more pronounced in the fifth century BC than in later times, for instance when the Ptolemies patronized researchers in Alexandria. See also Lloyd 2002:126–47.
30. Too often discussions of the prosecutions of the intellectuals Anaxagoras, Socrates, and a few others focus on the superstitions of the Athenians. It is much more likely that they were carefully orchestrated political moves to attack powerful leaders like Pericles through vulnerable allies. Such moves tap into a suspicious attitude among the people, but do not presuppose a spontaneous religious backlash. In the prosecution of Socrates the influential Anytus seems to have pulled the strings and used the unknown Meletus as a proxy to disguise the political character of the trial.
31. See Graham 2006:24–27 and n. 65.

hestiocentric to heliocentric, from circular orbits to epicycles to ellipses, from vortex dynamics to unmoved movers to gravitation to space-time deformations. According to the pessimistic induction argument all scientific theories are doomed to be refuted and replaced. But after two and a half millennia the indications are that at least some theories are here to stay. Here is a counter-example to pessimistic induction, as well as to the claim that paradigm shifts preclude the accumulation of knowledge. The present case study rather offers the basis for an optimistic induction argument.

The inevitable conclusion, however counter-intuitive it sounds in light of antirealist metaphysics and antiobjectivist hermeneutic theories, is that Anaxagoras' theory was irresistible because it was true. It produced its own empirical confirmation, of the sort that convinced Aristotle to employ it as a paradigm of scientific explanation. The theory accounted for all the relevant phenomena and satisfied all the predictions it made. This kind of success is rare, but it is the kind of success one should expect in a true theory.

Perhaps after a run of twenty-five centuries it is time to admit that Aristotle was right: some scientific theories just work, and explain their phenomena so completely that no further explanation is necessary. They may need to be adapted to new, more adequate theoretical contexts. But we see how a primitive explanation that we seem justified in calling a discovery can exercise a powerful influence in the history of science. Explain correctly the source of the moon's light, and you can explain the shape, nature, and path of some heavenly bodies. Organize these notions into a model, and you can explain eclipses. Account for all these things, and you can ask why the heavenly bodies follow the paths they do, what the shape of the earth is, what its size is, its distance from the sun and moon, the nature of heavenly bodies in general, the source of the sun's light. Some questions will have immediate answers, while others will take centuries to work out.

But it all starts with a simple explanation that replaces all others and generates its own evidence. It all starts, like a lot of science and social science, almost invisibly and undetectably, with a philosophical idea. An idea about light and dark that itself casts a long shadow in the history of human knowledge.

If the foregoing study is right, there are at least a few cases in which thinkers explained natural phenomena correctly from the beginnings of rational theorizing: heliophotism, antiphraxis, and further the sphericity of the earth (a theory proposed by Parmenides in the early fifth century but not vindicated until the fourth).[32] Let us not suppose that because these truths are now commonplace we can take them for granted. To the contrary, their acceptance marks the transition from a conjectural to a scientific cosmology, from stargazing to astronomy. Indeed, it is the acceptance of these foundational truths rather than the quest for mathematical models of planetary motion that marks the beginnings of Greek astronomy and hence of Western science. Most other ancient speculations have long since been consigned to the dustbin of history. But if a few ideas have proved themselves, metamorphosing from unfounded speculations to testable hypotheses to confirmed theories, surviving their inventors by many generations down to the present day, that is enough to show that philosophy is not an idle pastime, but a pathway to science; and that science itself is not a mere deceptive ordering of words, nor even a recent invention, but a tried and true pathway to knowledge of the natural world.

---

32. See above, ch. 3.2. The sphericity of the earth deserves a detailed study all its own. Like heliophotism it was proposed by Parmenides, apparently borrowed by Plato (*Phaedo* 108e ff.; but this is controversial—for a flat earth see Rosenmeyer 1956, Rosenmeyer 1959; for a hemispherical earth Morrison 1959; for a spherical earth Calder 1958; for a dodecahedron Couprie 2011:201–12) and supported by scientific proofs by Aristotle (*On the Heavens* II.14, 297a8 ff.), while in the third century BC the circumference of the earth was measured with reasonable accuracy by Eratosthenes.

APPENDIX 1

Anaxagoras in the Historiography of Science

The story embodied in this study makes Parmenides and Anaxagoras the heroes of early Greek astronomy, and indeed gives a chronological shift to the term "early Greek astronomy." It is Anaxagoras, however, who elaborated the insights of Parmenides, and who, according to Plutarch, made the case "in the clearest and boldest terms" for all to see.[1] And since scholars tend to ignore or downplay the cosmology of the former but not the latter, interpretations of Anaxagoras should provide a test case for an understanding of contributions of both philosophers. For purposes of comparison and contrast, let us review some important contributions to the historiography of science in general and astronomy in particular. This survey includes works with diverse aims and interests, from primers to monographs, so it is not surprising if all do not focus on the contributions of Anaxagoras; nevertheless the following survey will provide some sort of reference point for the present study.

William Whewell's groundbreaking *History of the Inductive Sciences* (3 vols., 1837, third edn. 1857) does not mention Anaxagoras in connection with astronomy. He sees heliophotism and the theory of eclipses as important advances, but mentions only Anaximander in the connection with the former and no one in connection with the latter development.[2]

---

1. *Nicias* 23.2–3 = 59A18.
2. Whewell 1857. His only reference to Anaxagoras in vol. 1 is p. 48 on his theory of matter. On heliophotism and antiphraxis, see 1:118–20.

APPENDIX 1

Paul Tannery, who has been called "the true founder of the modern history of science movement,"[3] expressed a deep admiration for Anaxagoras in *Pour l'histoire de la science hellène* (1887, 2nd edn., posthumous, 1930):

> [Anaxagoras] had the immortal honor of having given the first true, if incomplete, explanation of eclipses and of phases of the Moon; but we should note that this explanation was the hypothesis of a natural philosopher, not the result of an astronomer's observations. (287)

If he is not an astronomer, nevertheless he breaks new ground. In *Recherches sur l'histoire de l'astronomie ancienne* (1893) Tannery repeats his denial that what the Presocratics studied was astronomy.[4]

J. L. E. Dreyer's *History of the Planetary Systems from Thales to Kepler* (1906; reprinted as *A History of Astronomy from Thales to Kepler*, 1953) identifies Parmenides as recognizing heliophotism (21), Empedocles as correctly explaining solar eclipses (25), and Anaxagoras as accepting both heliophotism and antiphraxis (32).[5]

In a wide-ranging history of early meteorology, *Die meteorologischen Theorien des griechischen Altertums* (1907), Otto Gilbert sees Anaxagoras as offering an epoch-making advance—not, however, in his astronomy, but in his view that the heavens are composed of common earthy material, as illustrated by the stone of Aegospotami.[6]

Franz Boll wrote a detailed article on eclipses ("Finsternisse") in *Paulys Realencylopädie der classischen Atertumswissenschaft* (1909), which remains impressive to this day. In it he called into question the alleged contributions of Thales and Anaximenes, recognized Parmenides as the first to enunciate heliophotism (whatever was Parmenides' source, column 2342),[7] and saw Anaxagoras as having the best claim to the antiphraxis theory (2343).

---

3. Guerlac 1963:807. Tannery was very influential also in the history of Presocratic philosophy. Among other things, he is largely responsible for the inflated view of Pythagorean contributions to early thought.
4. "The period of the infancy of these principles [i.e., the period of the Presocratics] does not seem to me to belong at all to the history of astronomy" (Tannery 1893:29). "This need for naturalistic explanation [of the Presocratics] . . . did not, except for actual observations, suffice to surpass the limits which we assign to cosmography," because it did not use mathematics to account for the motion of the planets (30).
5. Dreyer 1953 is little changed from Dreyer 1906, but in its revival continued to be influential. Though Dreyer discusses Empedocles before Anaxagoras, he does not commit himself to whose work came first. He is skeptical of Thales' alleged achievements in this area (12–13). The Danish-born astronomer became the director of the Armagh Observatory in Northern Ireland and retired to Oxford.
6. Gilbert 1907:689 and n. 1. He sees the meteorite as a confirmation (Bestätigung) of the theory.
7. "Die Kenntnis der Beleuchtung des Mondes durch die Sonne steht absolut sicher . . . für Parmenides. . . ."

APPENDIX 1

One of the early histories of astronomy comes off well. Thomas Heath in his *Aristarchus of Samos: The Ancient Copernicus* (1913) observes:

> A great man of science, Anaxagoras enriched astronomy by one epoch-making discovery. This was nothing less than the discovery of the fact that the moon does not shine by its own light but receives its light from the sun. As a result, he was able to give ... the true explanation of eclipses. (78)

Although he gives perhaps too much credit to Anaxagoras (he thinks that Parmenides did not recognize heliophotism), Heath rightly grasps his importance for early astronomy.[8]

In his influential textbook *Early Greek Philosophy* (1892, 3rd edn. 1920),[9] John Burnet, who ascribes great importance to the Pythagorean tradition, observes:

> As to the moon's light and the cause of eclipses, it was natural that Anaxagoras should be credited at Athens with these discoveries. On the other hand, it seems very unlikely that they were made by a believer in a flat earth, and there is sufficient evidence that they are really Pythagorean. (272)

Thus Anaxagoras becomes a Footnote to Pythagoras.

In *Le système du monde* (10 vols., 1913–1959), Pierre Duhem follows Paul Tannery in giving an inflated importance to Pythagorean theories of the heavens, and he responds to Heath's *Aristarchus*. But he fails to take any notice of Anaxagoras' contributions to astronomy.[10]

W. A. Heidel in his study of the scientific achievements of the Greeks, *The Heroic Age of Science* (1933), does not mention Anaxagoras in connection with astronomy.

In a massive (and misnamed) study of ancient and medieval science, *An Introduction to the History of Science* (3 vols. in 5, 1927–1948), George Sarton gives Anaxagoras a half page of terse summary, which attributes heliophotism and antiphraxis to him.[11] In his *History of Science* (2 vols., 1952), Sarton allots Anaxagoras four long pages which provide his assessment of the Presocratic:

> [H]is scientific knowledge was not only meager but mostly wrong. His cosmologic views were forward, yet his astronomic knowledge was decidedly backward

---

8. On Parmenides and heliophotism, see Heath 1913:75–77. He also thinks that Anaxagoras could not have understood the phases of the moon because, since he held the earth was flat, he must also have viewed the moon and other heavenly bodies as flat (80–81). But as Dreyer 1906:32 had already pointed out, to explain the phases, he must have recognized the sphericity of the moon. Only a dubious scholium says the moon is flat (59A77).
9. The fourth edition of 1930 differs from the third only by a few corrections.
10. See Duhem 1954–1973, 1:5–27 for pre-Platonic theories.
11. Sarton 1927–1948, 1:86, leaving him out of discussions of astronomy elsewhere.

APPENDIX 1

as compared with that of the Pythagoreans. One cannot give him much credit for his explanation of the eclipses of the Sun and Moon by the interposition of Moon, Earth, or other bodies, because the explanation was not a novelty and because it was combined with crude ideas such as that the Earth and other planets are flat, that the Sun is larger than the Peloponnesos, and so on.[12]

Thus Sarton subscribes to the Footnotes to Pythagoras view.[13]

In his textbook *Greek Science* (2 vols, 1944, 1949, rev. edn. in 1 vol., 1961), Benjamin Farrington does not even discuss Anaxagoras' astronomical theory.

B. L. van der Waerden in his history of mathematics, *Science Awakening* (1954), writes that "Anaxagoras taught that the moon receives its light from the sun and he gave a correct account of solar and lunar eclipses." Nevertheless, "his influence on the development of astronomy was not wholly favorable. This was due chiefly to his view that the celestial bodies are inanimate objects which do not move according to mathematical laws, but are dragged along by the vortical motion of the ether" (128).

In his study *The Physical World of the Greeks* (1956), Shmuel Sambursky notes, "Most of the Greek astronomers accepted Anaxagoras' theory that the moon has no light of its own, but that 'it is the sun that endows the moon with its brilliance' [B18]." He goes on to say, "While it is true that ... Anaxagoras' astronomical ideas were primitive and not original [n.b.], this in no way detracts from his importance as a pioneer of the concept of the unity of celestial and terrestrial phenomena" (24). Sambursky does not connect Anaxagoras with the explanation of eclipses.

Historian of astronomy Otto Neugebauer does not mention Anaxagoras either in his *Exact Sciences in Antiquity* (2nd edn., 1957) or his *History of Ancient Mathematical Astronomy* (3 vols., 1975).[14]

Similarly, in *The Copernican Revolution* (1957), which traces astronomical theories back to the Greeks, Thomas Kuhn does not mention Anaxagoras.

E. J. Dijksterhuis in *The Mechanization of the World Picture* (1950, English trans. 1961), discusses only Anaxagoras' theory of elements.

In *A History of Astronomy* (1951, English trans. 1961), Anton Pannekoek attributes to Anaxagoras the first clear statement of heliophotism and of antiphraxis for lunar eclipses, but he does not mention solar eclipses.[15]

In *The Fabric of the Heavens* (1961), Stephen Toulmin and June Goodfield recognize Anaxagoras as giving the right account of eclipses, but do not tie that account in to the further development of astronomy.[16]

---

12. Sarton 1952, 1:243.
13. For this view, see ch. 1.3; for problems with it, see ch. 3.2.
14. In Neugebauer 1957, ch. 6, he jumps from Babylonian to Hellenistic astronomy. Neugebauer 1975, vol. 2, gives a cursory overview of early Greek contributions.
15. Pannekoek 1961:100.
16. Toulmin and Goodfield 1961:68–69, 71–72.

APPENDIX 1

Giorgio de Santillana discusses a broad range of topics in *The Origins of Scientific Thought: From Anaximander to Proclus, 600 BC to 300 AD* (1961), but in his discussion of astronomy (chs. 15–16, cf. 12) he has nothing to say about anyone before Eudoxus.

Marshall Clagett widely used textbook, *Greek Science in Antiquity* (1963), does not mention Anaxagoras in its chapter on astronomy.[17]

Daniel E. Gershenson and Daniel E. Greenberg provide a lengthy sourcebook on Anaxagoras in *Anaxagoras and the Birth of Physics* (1964a, digested in 1964b), specifically addressed to the Presocratic's scientific contributions. The section on astronomy (1964a:41–43) comments briefly on cosmological issues, but says almost nothing about astronomical problems.

In his *History of Greek Philosophy* (vol. 2, 1965), W. K. C. Guthrie denies heliophotism to Parmenides and attributes both that theory and antiphraxis to Anaxagoras, on the authority of Plato and Plutarch.[18]

I began the book by looking at some comments in D. R. Dicks' *Early Greek Astronomy to Aristotle* (1970). Dicks grants that Anaxagoras accepted heliophotism and antiphraxis (57–59), and concedes that "astronomical thought in the latter part of the fifth century BC was beginning to move away from the speculative theorizing of the earlier thinkers towards a more empirical attitude . . ." (59). This conclusion, however, only provides the prelude to his Unfounded Speculation argument, discussed above (ch. 1.1).

At about the same time, by contrast, Kurt von Fritz gives an unqualifiedly positive assessment of Anaxagoras in his monograph *Der Ursprung der Wissenschaft bei den Griechen* (1971b).[19] After a careful reconstruction of his astronomical theory, he concludes:

> To summarize, it turns out that cosmological astronomy made astonishing progress toward modern views through Anaxagoras. Most astonishing is the knowledge that the stars consist not of light matter, that by nature strives upward or can "float" in the air, but of solid and heavy stone, and that they are hindered from falling down to earth by their motion; in a certain sense an anticipation of the Cartesian vortex theory. . . .[20] The second is an explanation that, if not totally correct, at least closely approximates the correct one, of solar and lunar eclipses as well as of the phases of the moon,[21] based on the knowledge that the moon gets its light from the sun. In conjunction with this

---

17. It makes only one mention of Anaxagoras, concerning his theory of *nous* (p. 52).
18. Guthrie 1962–1981, 2:66, 306–7.
19. This is an expanded version of von Fritz 1961, included in the collection von Fritz 1971a.
20. Here von Fritz claims that this conception did not catch on in antiquity, which seems to me false: most Presocratics after him follow this conception, as well as the Epicureans.
21. Von Fritz thinks that Anaxagoras may have taken the moon to be a flat disk, citing incorrectly 59A55 (perhaps he means to cite the scholium to Apollonius of Rhodes, A77; see 153 and n. 301; but this testimony is of uncertain pedigree and value).

APPENDIX 1

emerges the essentially correct ordering of the distances of the main heavenly bodies from the earth, and finally the correct explanation of the "face" of the moon. (154)[22]

Von Fritz is one of the few scholars who does justice to Anaxagoras. Unfortunately, his study seems to have had little influence on subsequent research in the area.

One of the most influential historians of Greek science of recent times is Geoffrey Lloyd. His introductory *Early Greek Science: Thales to Aristotle* (1970) treats Anaxagoras only in relation to the problem of change, not astronomy.[23] His study of the methods and social setting of science, *Magic, Reason and Experience: Studies in the Origin and Development of Greek Science* (1979), discusses early astronomy in a chapter on The Development of Empirical Research:

> Various authors are credited with knowing—or themselves indicate that they know—that the moon shines with reflected light, that the Morning Star and the Evening Star are one and the same body, and that the interposition of the moon and of the earth causes eclipses of the sun and moon respectively. Although our sources disagree about who precisely "discovered" these facts (that is who among the Greeks first recognized them), we may take it that some at least of those who were engaged in astronomical speculation in the late fifth and early fourth centuries were well aware of them.... (170–71)[24]

Thus Lloyd grants that there was empirical progress among the Presocratics, though he does not venture to sort out the competing claims. He concludes his book by saying,

> Although eventually Greek scientists produced lasting (if often elementary) results in the areas of astronomy [etc.] ... Greek science down to Aristotle is more notable for its achievements in second-order inquiries, in epistemology, logic, methodology and philosophy of science.... (266)

Claire Préaux provides a monograph on the moon in ancient literature, philosophy, and science, *La lune dans la pensée grecque* (1973), which defends Anaxagoras as the discoverer of the antiphraxis theory.[25] Parmenides recognized the moon as always turning its luminous face to the sun, but did not necessarily espouse

---

22. For his whole treatment of Anaxagoras' astronomy, see 151–55.
23. Ch. 4, The Problem of Change, 36–49; cf. ch. 7, Fourth-Century Astronomy. There is no systematic treatment of fifth-century astronomy.
24. 270 nn. 228–230 give the evidence for various philosophers holding the views, including Anaxagoras.
25. Préaux 1973:156–87.

APPENDIX 1

heliophotism (172). Empedocles wrote after Anaxagoras and adopted his astronomical advances (175).

In his *Early Physics and Astronomy: A Historical Introduction* (1974, revised edn., 1993), Olaf Pedersen briefly recognizes the contributions of Parmenides to heliophotism and Anaxagoras to the explanation of eclipses (35).

In his study of Plato's science, *Plato's Universe* (1975), Gregory Vlastos does not talk about Anaxagoras' contributions to astronomy, though he mentions him a number of times.

In *The Beginnings of Western Science* (1992), David Lindberg recognizes Anaxagoras and Empedocles as giving the correct account of eclipses, but he does not link this with his account of astronomy.[26]

Hugh Thurston, *Early Astronomy* (1994), ascribes heliophotism and antiphraxis to Anaxagoras in a single sentence (110–11).

In a journal article, "Wer entdeckte die Quelle des Mondlichts?" (1995), Georg Wöhrle argues for what I take to be the correct answer to the problem of heliophotism, that Parmenides first enunciated the theory, followed by Anaxagoras and Empedocles, who gave clearer and more systematic accounts of it.[27]

Robert Wilson, *Astronomy Through the Ages* (1997), does not mention Anaxagoras.

In *The History and Practice of Ancient Astronomy* (1998), James Evans provides a rich study of the lore and techniques of ancient astronomers from Mesopotamia and Greece. He gives a good explanation of Anaxagoras and his accomplishments (46). But he tends to separate the philosophical (and the literary) traditions from the scientific and to see only mathematical constructs as scientific, although he does allow for influence among traditions (17–25).

As noted above (ch. 3.4), Karl Popper in the posthumous collection of essays, *The World of Parmenides* (1998), argues vigorously that Parmenides' doctrine of heliophotism marked an important advance. He does not, however, follow the influence of the idea to Anaxagoras or other figures, or to the history of astronomy in general. He is rather interested in the impact of the idea on Parmenides' philosophical and cosmological theories (which he regards as closely related).

In *Greek Science* (1999), T. E. Rihll mentions Anaxagoras twice, but not in connection with astronomy (ch. 4).

Edward Grant in *A History of Natural Philosophy* (2007) does not mentions Anaxagoras at all.

In *Conceptions of Cosmology* (2007), Helge Kragh mentions Empedocles as recognizing that the moon gets its light from the sun and that the earth causes lunar eclipses, but he does not attribute this view to Anaxagoras (14).

Most recently, Dirk Couprie in his *Heaven and Earth in Ancient Greek Cosmology* (2011), which focuses on cosmology to the exclusion of astronomy (but does have

---

26. Lindberg 1992:26, cf. 88 ff.
27. See also Graham 2002.

APPENDIX 1

much to say about astronomy in my broader sense), attributes to Anaxagoras heliophotism (which Couprie believes he originated) and antiphraxis.[28]

We see, then, that the view that Anaxagoras developed heliophotism and introduced the antiphraxis doctrine is not new. It was developed by Tannery, Boll, and Heath in the late nineteenth and early twentieth centuries—a French, a German, and a British champion of an international interpretation. Yet the continued importance of Thales and the rising status of Pythagoras and his "school" tended to make Anaxagoras a Footnote to Thales or, more often, to Pythagoras, while strict scientific interpreters have often reduced all Presocratic theories to Unfounded Speculations. (I presume that some at least of those historians who fail to mention Anaxagoras' astronomy subscribe to the last view.) Even those who recognize Anaxagoras for heliophotism and antiphraxis often find his contributions disappointing because they are too elementary or not mathematical enough. Many classify Anaxagoras' knowledge as cosmology but not astronomy, and see no link between the former and latter domains. This to me is the biggest mistake of all: to give him credit for a significant advance, but then continue to tell the story of astronomy as if that advance did not matter to later developments. While many recent scholars recognize Anaxagoras' knowledge of lunar light and eclipses, few appreciate its importance for the further elaboration of astronomy.

---

28. Couprie 2011:176–78 sees Anaxagoras as the first to view heavenly bodies as "fiery masses of stone." He holds that Anaxagoras' moon is disk-shaped, which makes his understanding of the phases of the moon impossible.

APPENDIX 2

Science and History

In chapter 1, I spent some time dealing with antirealist views of science and antiobjectivist views of history of science.[1] I did not offer a position of my own to ground the reconstruction of early Greek science I subsequently presented. It is time to offer my own manifesto. Yes, there is such a thing as Science (which in the present polemic I shall dignify with a capital letter) in a strong sense (to be specified presently). It is true that there is no single scientific method, no body of facts, and no shared theory that timelessly characterizes the practice. What Science offers is a systematic way of understanding nature. Now if nature did not exhibit a high degree of regularity, and if human beings were not capable of apprehending complex regularities, there would be no possibility of Science. But there is inductive evidence that both these preconditions obtain. Hence, Science is in principle possible, and the present improved quality of life provides historical evidence that actual scientific understanding at least sometimes comes about.

I take it that human beings learn at least some of what they know through sense perception of the world around them. Further, human beings are social animals who

---

1. Here I continue to respond to the papers in *Isis* 83 (1992), The "Cultures of Ancient Science: Some Historical Reflections," which advocate a new approach to the historiography of science (see above, ch. 1.6, 1.7). For instance, von Staden (1992) finds that claims of "affinity" between ancient and modern projects are "insidious" in promoting "the unspoken assumption that such affinities are especially, or even uniquely, worthy of our historiographical efforts" (583). Such an approach would rule out, a priori it seems, any study that finds a continuity between ancient and modern science. (Von Staden backpedals a bit, 584, but his accent is clearly on the dangers of affinity.) I take it that this general approach represents an important movement in recent historiography of science.

necessarily (given their individual limitations and need of long-term nurture in childhood) live in communities, who communicate verbally, and join in community projects. Science, as a first approximation, is a socially organized pursuit of knowledge of the natural world that can in principle extend the capabilities of the individual learner. Science is thus the product of culture, but it becomes efficacious, or properly comes to exist, when inquirers learn to put questions to nature in an appropriate way and to evaluate the responses. There is no scientific inquiry without an appropriate cultural framework, and likewise none without some mechanism for putting the question to nature. Thus culture (Nomos) asks the questions, nature (Physis) gives the answers. "Reason" as Immanuel Kant perceptively explained, "must approach nature in order to be taught by it. It must not, however, do so in the character of a pupil who listens to everything that the teacher chooses to say, but of an appointed judge who compels the witnesses to answer questions which he has himself formulated."[2] Accordingly, a student of the history of science must be familiar with both culture and nature to understand how, on the one hand, the questions were framed, and how, on the other, they were answered—and the answers were understood.

I tentatively offer the following as a working definition of Science:

S. Science is a) a systematic study of the natural world, b) using an accepted theory and methodology, c) allowing for open inquiry within (b), d) permitting elaboration and revision of (b), e) based on empirical evidence.

Science needs to be a systematic rather than haphazard confrontation with nature (a).[3] It needs to involve a communal effort in which there is some shared background by inquirers, as with a Kuhnian paradigm, rather than a mere set of conjectures by an individual (as seems to have characterized much early Greek philosophy of nature), but with enough content to guide practice (b). The system has to be in principle corrigible by inquirers so as to be adaptable to experience rather than being dogmatic and inflexible (c, d). And above all, it has to be responsive in some significant way to evidence obtained through inquiry (e). I take it that a dogmatic system could become scientific if it were subjected to testing,[4] while a scientific system could become ossified into a non-scientific dogma if inquiry were abandoned.[5] Any practice that

---

2. *Critique of Pure Reason* Bxiii, trans. N. K. Smith.
3. The notion of "nature" or "the natural world" is itself theory-laden and historically conditioned. But at least from early Greek times there has been some such notion or family of notions. See Naddaf 2005; Lloyd 1991b.
4. Which may have happened at some point among Presocratic theories, and did, as I argue in this work.
5. Which may have happened among late scholastic versions of Aristotelian science, which in the hands of Aristotle included empirical researches into, e.g., biology and astronomy. For a recent example of dogmatism from anthropology, see Pinker 2002:115–19.

embodies the points above could count as a science. Built in to this definition is at least the possibility of a progressive understanding of the world, which, *pace* some historians of science,[6] I take to be one of the chief virtues and perhaps the most distinctive feature of Science.

Nothing in **S** makes reference to any specific method, time, place, or culture. It does not specify the use of instruments, the need for experiments, or the requirement of quantitative measurements or mathematical formulations. It does not even exclude an appeal to religious beliefs or mythological assumptions, so long as the stated conditions are met. Nor does it presuppose formal record keeping nor consequently literacy.[7] Further, it remains neutral as to whether scientific practice is to be understood as inductive, abductive, hypothetico-deductive, or some mixture of these logical approaches.

Whether any particular practice in any particular culture counts as a science should be understood as an empirical question. I believe some ancient Greek practices, including astronomy, meet these conditions. I believe some non-Greek practices, including Babylonian astronomy-cum-astrology, also meet them. Hence **S** does not, to my mind, exhibit an undesirable cultural bias. Further, it is, I think, broad enough to allow for multiple instantiations within the same time, culture, and domain; for example, rival schools of medicine might fulfill these conditions. Now I am not sure **S** is either broad enough or narrow enough to meet all explanatory needs. But I do believe that unless we acknowledge *some* normative criterion in our research, the history of science is likely to collapse into a mere cultural study of belief systems, exuding an air of importance and relevance it no longer deserves. My claim is that instances of Science are possible and sometimes actual in human history, perhaps an inevitable outcome of collective human problem-solving efforts. When any culture arrives at a scientific practice, the occasion is worthy of study and even offers a cause for celebration. There is no one science, no one scientific method, and no one scientific tradition. Nevertheless, some cultures may, in fact, arrive at scientific understandings and scientific practices before others and influence them. In many cases our lack of research may keep us from appreciating the achievements of some cultures, a lack that should be corrected by appropriate study. But the history of science is not a zero-sum game: to acknowledge an achievement in one culture is not ipso facto to disparage an achievement in another culture.

6. "[T]he papers that follow implicitly criticize an earlier historiography of science that presupposes progress as a characteristic of science and objectivity as a characteristic of its historians" (Rochberg 1992:551). See also her remarks on the alleged vice of "Continuism" (552), specifically related to astronomy, seconding von Staden (1992:586 et passim).

7. Of course without written records scientific practices will be inaccessible to historians. But I do not want to rule out a priori the possibility of scientific practices existing in preliterate societies. A strong oral tradition could, in principle, lead to a later written record. We may have such a case with Thales; see ch. 2.2.1. Archaeoastronomy tries to reconstruct prehistoric astronomy of preliterate cultures on the basis of architectural alignments and related archaeological finds.

APPENDIX 2

It would be unfortunate indeed if, in light of a legitimate concern for non-Greek cultures, we were to abjure the study of one of the cultures that we know made significant scientific advances. Research into Greek science, even when it establishes a direct line of succession to modern science, does not constitute an assault on non-Greek science. Furthermore, an affirmation of Science in general is not a denigration of any particular culture. Pingree (1992) complains of Hellenophilia that it maintains that (1) the Greeks invented science and (2) they discovered a scientific way to truth (see above, ch. 1.6). In light of **S** let me answer: yes, the Greeks did invent science (or more precisely sciences) and they did discover a scientific way to truth—and so did other cultures. I should think that the history of science just is the study of such cases, not the attempt to re-envisage science in such a way that every culture comes out a winner no matter what. In our present state of knowledge, it appears that influence between Greek and Mesopotamian scientific cultures was bi-directional, and that Mesopotamian science was clearly superior to Greek science in every way until the fifth century BC and in many ways after that date. So I see no reason for the cultural defensiveness that characterizes some essays in the new historiography.

There seem to be at least three conceptions of the history of science in play in recent discussions.[8] First, what we may call the positivist view, according to which there is a straightforward scientific method that one can learn, apply to problems, and solve them; to reconstruct a historical discovery is just to rethink the problem by applying scientific method. Second, what I shall call the paradigmatic view first developed by Kuhn that recognizes methodological progress within a paradigm or research program, but that also recognizes times of crisis in which a radically new approach is required to solve intransigent problems ("anomalies"); crises lead to times of revolution in which new paradigms win out and redefine the scientific program. Finally, there is what I shall call a descriptivist view that seems to count as science whatever anyone takes to be an explanation of phenomena, whether "perceived or imaginary."[9] In this view every culture has a science and every science is equally worthy of study by historians of science.

I want to identify my own view with a Kuhnian approach, minus the antirealism. That is, I think one can accept much of the phenomenology of science Kuhn presents without accepting his claims that there can be no rational grounds for preferring one paradigm over another.[10] What his view offers is a way to acknowledge normal science along with revolutionary science, and to recognize that scientific method is not

---

8. I omit earlier conceptions such as the Aristotelian and Hegelian, both of which are teleological in an important way. Also, I am shamelessly simplifying the options here for the present context.
9. "[S]cience is a systematic explanation of perceived or imaginary phenomena, or else is based on such an explanation" (Pingree 1992:559).
10. I think § XIII "Progress Through Revolutions" (the title is of course misleading) of Kuhn 1996 [1962] overlooks the ways in which one paradigm can improve on another, perhaps because of the antirealist epistemology lurking in the background.

unitary nor progress linear. There are times at which existing theories and methodologies fail to solve the problems they face, and radical new approaches are explored and sometimes adopted. There is not, then, a steady and predictable march of science into the future, and scientific method is not an algorithm that can be applied willy-nilly to any problem. Yet neither is science a self-contained game, nor is it a mere cultural practice. Crucially, Kuhn recognizes a normative dimension to science, embodied in the paradigm, and that is what I think is fundamental to Science. Certainly Science *aims* at a progressive understanding of the world, and I believe history demonstrates that Science succeeds, if not in every case, then overall; and not only within normal science but also across revolutionary episodes.[11]

The descriptivists seem to want to abandon any normative function in favor of whatever a culture or community believes in. But if we make the conditions for science too loose, then we cannot distinguish science from other kinds of knowledge or lore—not only pseudo-science, but philosophy, mathematics, theology, and, yes, divination. The descriptivists seem to be motivated in part by multicultural ideals and a touch of Hellenophobia, with a desire to do away with "ethnocentric rubbish."[12] But the methodological question, it seems to me, is independent of and prior to the cultural question. We cannot determine the nature of science by surveying cultures to see what they believe in; we must determine it on the basis of what it is for us.[13] The history of science has a certain prestige precisely because science has a certain prestige. Now a historian can redefine science as anything he pleases, such as "a systematic explanation of perceived or imaginary phenomena,"[14] but this does not guarantee that he is studying the history of science if he is studying, for example, the systematic explanation of religious phenomena. There would be nothing wrong with studying the history of knowledge, for instance. But such a pursuit would count as history of science only if the subject matter were recognizably scientific.

Consider an analogy. Historians of philosophy do not claim to consider all interesting ideas or conceptions, only theories they recognize as philosophically valuable. Now to define philosophy is considerably more difficult than to define science; yet historians of philosophy (who are a kind of philosopher) can recognize it when they see it. There is a kind of normativity to philosophy, which aims at truth rather than persuasion, as Plato maintained.[15] But this does not preclude historians from studying ideas and conceptions that fall short of being philosophically compelling.

---

11. The political models of revolution which Kuhn draws on tend to suggest that revolutionary changes constitute quantum leaps in social or economic or political organization.
12. Pingree (1992:555).
13. "The history of science can never be *purely* descriptive any more than the practice of science can" (Lloyd 1992:565, his italics). "We must be clear, then, as a first principle, that the history of science is inevitably evaluative: it always presupposes a conceptual framework and a methodology" (566).
14. See n. 9 above.
15. *Gorgias* 454e–455a, 482b–c.

APPENDIX 2

We call such researchers historians of ideas and intellectual historians. Their fields of study are perfectly respectable. But they are not philosophy. If historians of philosophy became historians of ideas, philosophers would soon lose interest in their work; and if historians of science became historians of folklore, for instance, scientists and students of science would lose interest in their work likewise.

In the end, I think we need to bring a robust notion of science to the history of science. It should not be so narrow as to allow, for instance, only cases of twenty-first century science to count, or be so broad, on the other hand, as to allow any kind of lore or mythology. But without some working hypothesis about what science is we cannot even ask whether anyone in antiquity was engaged in scientific practices. And without some interest in the successes of modern science, there is not much reason to even ask the question.

BIBLIOGRAPHY

Aaboe, Asger. 1972. "Remarks on the Theoretical Treatment of Eclipses in Antiquity." *Journal for the History of Astronomy* 3: 105–18.
Ancient Near Eastern Texts. 1969. 3d ed. Edited by James B. Pritchard. Princeton: Princeton University Press.
Bailey, Cyril. 1928. *The Greek Atomists and Epicurus*. Oxford: Clarendon Press.
Baldry, H.C. 1928. "Embryological Analogies in Pre-Socratic Cosmogony." *Classical Quarterly* 26: 27–34.
Barnes, Jonathan. 1982 [1979]. *The Presocratic Philosophers*. Rev. ed. London: Routledge & Kegan Paul.
Beatty, Mario. 1997/1998. "On the Source of the Moon's Light in Ancient Egypt." *ANKH: Révue d'Égyptologie et des Civilisations Africaines* 6/7: 162–77.
Bernal, Martin. 1992. "Animadversions on the Origins of Western Science." *Isis* 83: 596–709.
Bickerman, E.J. 1980. *Chronology of the Ancient World*. Ithaca, NY: Cornell University Press.
Bicknell, Peter J. 1967a. "Xenophanes' Account of Solar Eclipses." *Eranos* 65: 73–77.
———. 1967b. "A Note on Xenophanes' Astrophysics." *Acta Classica* 10: 135–36.
———. 1968a. "The Planet Mesonux." *Apeiron* 2: 10–12.
———. 1968b. "Did Anaxagoras Observe a Sunspot in 467 B.C.?" *Isis* 59: 87–90.
———. 1969. "Anaximenes' Astronomy." *Acta Classica*, 12: 53–85.
Blanche, Lenis. 1968. "L'éclipse de Thalès et ses problèmes." *Revue Philosophique de la France et de l'Étranger* 158: 153–99.
Bodnár, István M. 1988. "Anaximander's Rings." *Classical Quarterly* 38: 49–51.
———. 1992. "Anaximander on the Stability of the Earth." *Phronesis* 37: 336–42.
Bodnár, István, and William W. Fortenbaugh, eds. 2002. *Eudemus of Rhodes*. New Brunswick, NJ: Transaction Publishers.

BIBLIOGRAPHY

Boll, Franz. 1909. "Finsternisse." In *Paulys Realencyclopädie der classischen Altertumswissenschaft*. Edited by Georg Wissowa. Vol. 6, cols. 2329–64. Stuttgart: J.B. Metzler.

Bollack, Jean. 1965–1969. *Empédocle*. 3 vols. Paris: Les Éditions de Minuit.

———. 1990. "La cosmologie parménidéenne de Parménide." In *Herméneutique et ontologie: mélanges en hommage à Pierre Aubenque*, edited by Rémi Brague and Jean-Francois Courtine, 17–53. Paris: Presses Universitaires de France.

———. 2006. *Parménide: de l'étant au monde*. Lagrasse: Éditions Verdier.

Bonitz, Hermann. 1848–1849. *Aristotelis Metaphysica*. 2 vols. Bonn: Marcus. Repr. Hildesheim: Olms, 1960.

Bonneau, Danielle. 1964. *La crue du Nil: divinité égyptienne à travers mille ans d'histoire (332 av.-641 apr. J.-C.)*. Paris: Librairie C. Klincksieck.

Bowen, Alan C. 2002. "Eudemus' History of Early Greek Astronomy: Two Hypotheses." In Bodnár and Fortenbaugh 2002, 307–22.

Bowen, Alan C., and Bernard R. Goldstein. 1988. "Meton of Athens and Astronomy in the Late Fifth Century B.C." In Leichty et al. 1988, 39–81.

———. 1994. "Aristarchus, Thales, and Heraclitus on Solar Eclipses: An Astronomical Commentary on P.Oxy. 53.3710 Cols. 2.33–3.19." *Physis* 31: 689–729.

Bowen, Alan C., and Robert C. Todd. 2008. "Kleomēdēs." In *The Encyclopedia of Ancient Natural Scientists*, edited by Paul T. Keyser and Georgia L. Irby-Massie, 479–80. London: Routledge.

Boyd, Richard N. 1984. "The Current Status of Scientific Realism." In Leplin 1984, 40–82.

Burch, George Bosworth. 1954. "The Counter-Earth." *Osiris* 11: 267–94.

Burkert, Walter. 1972 [1962]. *Lore and Science in Ancient Pythagoreanism*. Translated by E.L. Minar Jr. Cambridge: Harvard University Press.

———. 1992. *The Orientalizing Revolution*. Cambridge: Cambridge University Press.

Burnet, John. 1930 [1892]. *Early Greek Philosophy*. 4th ed. London: Adam & Charles Black.

Calder, William M., III. 1958. "The Spherical Earth in Plato's *Phaedo*." *Phronesis* 3: 121–25.

Cappelletti, Angel G. 1979. "Notas para una biografía de Anaxágoras." *Diálogos* 14: 7–28.

Casson, Lionel. 1971. *Ships and Seamanship in the Ancient World*. Princeton: Princeton University Press.

Cerri, Giovanni. 1999. *Parmenide: poema sulla natura*. Milan: Rizzoli.

Chakravartty, Anjan. 2007. *A Metaphysics for Scientific Realism: Knowing the Unobservable*. Cambridge: Cambridge University Press.

Cherniss, Harold. 1935. *Aristotle's Criticism of Presocratic Philosophy*. Baltimore: The Johns Hopkins University Press.

Cherniss, Harold, and William C. Hembold, trans. 1957. *Plutarch: Moralia*, vol. 12. Loeb Classical Library. Cambridge: Harvard University Press.

Citroni Marchetti, Sandra. 2007. "Plinio il vecchio, Anassagora, e le pietre cadute dal sole." *Studi Italiani di Filologia Classica* Ser. 4, 5: 125–55.

BIBLIOGRAPHY

Clagett, Marshall. 1963 [1955]. *Greek Science in Antiquity.* 2d ed. London: Collier.
Classen, C. Joachim. 1965. "Bemerkungen zu zwei griechischen 'Philosophiehistorikern.'" *Philologus,* 109: 175–81.
Collingwood, R.G. 1946. *The Idea of History.* Oxford: Clarendon Press.
Conche, Marcel. 1986. *Héraclite: Fragments.* Paris: Presses Universitaires de France.
———. 1999 [1996]. *Parménide: le poème: fragments.* 2d ed. Paris: Presses Universitaires de France.
Cordero, Néstor-Luis et al. 2008. *Eleatica 2006: Parmenide scienzato?* Edited by Livio Rossetti and Flavia Marcacci. Sankt Augustin, Germany: Academia Verlag.
Connor, W.R. 1993. "The *histor* in History." In *Nomodeiktes: Studies in Honor of Martin Ostwald,* edited by Ralph M. Rosen and Joseph Farrell, 3–15. Ann Arbor: University of Michigan Press.
Cornford, F.M. 1937. *Plato's Cosmology.* London: Routledge & Kegan Paul.
———. 1939. *Plato and Parmenides.* London: Routledge & Kegan Paul.
Couprie, Dirk L. 1995. "The Visualization of Anaximander's Astronomy." *Apeiron* 28: 159–81.
———. 2006. "Anaxagoras und die Grösse der Sonne." *Hyperboreus* 12: 55–76.
———. 2009a. "Problems with Anaximander's Numbers." *Apeiron* 42: 167–83.
———. 2009b. "The Tilting of the Heavens in Presocratic Cosmology." *Apeiron* 42: 259–74.
———. 2011. *Heaven and Earth in Ancient Greek Cosmology: From Thales to Heraclides Ponticus.* New York: Springer.
Couprie, Dirk L., Robert Hahn, and Gerard Naddaf. 2003. *Anaximander in Context.* Albany: State University of New York Press.
Coxon, A.H. 1986. *The Fragments of Parmenides.* Assen, Netherlands: Van Gorcum.
Curd, Patricia. 1998. *The Legacy of Parmenides.* Princeton: Princeton University Press.
———. 2007. *Anaxagoras of Clazomenae: Fragments and Testimonia.* Toronto: University of Toronto Press.
Curd, Patricia, and Daniel W. Graham, eds. 2008. *The Oxford Handbook of Presocratic Philosophy.* New York: Oxford University Press.
Davison, J.A. 1953. "Protagoras, Democritus, and Anaxagoras." *Classical Quarterly,* n.s., 3: 33–45.
Devitt, Michael. 2010. *Putting Metaphysics First: Essays on Metaphysics and Epistemology.* Oxford: Oxford University Press.
Dicks, D.R. 1959. "Thales." *Classical Quarterly,* n.s., 9: 294–309.
———. 1966. "Solstices, Equinoxes, and the Presocratics." *Journal of Hellenic Studies* 86: 26–40.
———. 1970. *Early Greek Astronomy to Aristotle.* Ithaca, NY: Cornell University Press.
Diels, Hermann. 1879. *Doxographi graeci.* Berlin; reprint Berlin: W. de Gruyter, 1965.
———. 1897. *Parmenides Lehrgedicht.* Berlin: Georg Reimer.
Dijksterhuis, E.J. 1961 [1950]. *The Mechanization of the World Picture.* Translated by C. Dikshoorn. Oxford: Oxford University Press.

BIBLIOGRAPHY

Drake, Stillman, trans. 1967 [1953]. *Galileo Galilei: Dialogue Concerning the Two Chief World Systems—Ptolemaic and Copernican.* 2d ed. Berkeley: University of California Press.

Dreyer, J.L.E. 1953 [1906]. *A History of Astronomy from Thales to Kepler,* 2d ed. Edited by W.H. Stahl. New York: Dover.

Duhem, Pierre. 1954–1973 [1913–]. *Le système du monde: histoire des doctrines cosmologiques de Platon à Copernic.* 10 vols. Paris: Hermann.

———. 1969 [1908]. *To Save the Phenomena: An Essay on the Idea of Physical Theory from Plato to Galileo.* Translated by Edmund Doland and Chaninah Maschler. Chicago: University of Chicago Press.

Edelstein, L., and I.G. Kidd. 1972–1988. *Posidonius.* 3 vols. Cambridge: Cambridge University Press.

Evans, James. 1998. *The History and Practice of Ancient Astronomy.* New York: Oxford University Press.

Farrington, Benjamin. 1961. *Greek Science.* Baltimore: Penguin.

Fehling, Detlev. 1985. "Das Problem der Geschichte des griechischen Weltmodells vor Aristoteles." *Rheinisches Museum* 128: 195–231.

———. 1994. *Materie und Weltbau in der Zeit der frühen Vorsokratiker.* Innsbruck: Institut für Sprachwissenschaft der Universität Innsbruck.

Ferguson, John. 1971. "Dinos." *Phronesis* 16: 97–115.

Fine, Arthur. 1984. "The Natural Ontological Attitude." In Leplin 1984, 83–107.

Finkelberg, Aryeh. 1997. "Xenophanes' Physics, Parmenides' Doxa and Empedocles' Theory of Cosmogonical Mixture." *Hermes,* 125: 1–16.

Floyd, Edwin D. 1990. "The Sources of Greek *histōr* 'Judge, Witness.'" *Glotta* 68: 157–66.

Frank, Erich. 1923. *Plato und die sogenannten Pythagoreer: Ein Kapitel aus der Geschichte der griechischen Geistes.* Halle: Max Niemeyer.

Frankfort, Henri, H. A. Frankfort, John A. Wilson, and Thorkild Jacobsen. 1949. *Before Philosophy: The Intellectual Adventure of Ancient Man.* Baltimore: Penguin.

Furley, David J. 1987. *The Greek Cosmologists,* Vol. 1: *The Formation of the Atomic Theory and Its Earliest Critics.* Cambridge: Cambridge University Press.

———. 1989. *Cosmic Problems.* Cambridge: Cambridge University Press.

Gamow, George. 1964. *A Star Called the Sun.* New York: Viking Press.

Gershenson, Daniel E., and Daniel A. Greenberg. 1964a. *Anaxagoras and the Birth of Physics.* New York: Blaisdell.

———. 1964b. *Anaxagoras and the Birth of Scientific Method.* New York: Blaisdell.

Gianvittorio, Laura. 2010. *Il discorso di Eraclito: un modello semantico e cosmologico nel passaggio dall' oralità alla scrittura.* Hildesheim: Georg Olms.

Gilbert, Otto. 1907. *Die meteorologischen Theorien des griechischen Altertums.* Leipzig: B.G. Teubner.

Goldin, Owen. 1996. *Explaining an Eclipse: Aristotle's Posterior Analytics 2.1-10.* Ann Arbor: University of Michigan Press.

Goldstein, Bernard R., and Alan C. Bowen. 1983. "A New View of Early Greek Astronomy." *Isis* 74: 330–40.

Görgemanns, Herwig. 1970. *Untersuchungen zu Plutarchs Dialog De facie in orbe lunae.* Heidelberg: Carl Winter.

Graham, Daniel W. 1988. "Symmetry in the Empedoclean Cycle." *Classical Quarterly* 38: 297–312.

———. 1991. "Socrates, the Craft Analogy, and Science." *Apeiron* 24: 1–24.

———. 2002. "La lumière de la lune dans la pensée grecque archaïque." In Laks and Louguet 2002, 351–80.

———. 2003a. "A New Look at Anaximenes." *History of Philosophy Quarterly* 20: 1–20.

———. 2003b. "Philosophy on the Nile: Herodotus and Ionian Research." *Apeiron* 36: 291–310.

———. 2006. *Explaining the Cosmos: The Ionian Tradition of Scientific Philosophy.* Princeton: Princeton University Press.

———. 2008. "Socrates on Samos." *Classical Quarterly* 58: 308–13.

———. 2013. "Anaxagoras and the Comet." *Ancient Philosophy* 33: 1–18.

Graham, Daniel W., and Eric Hintz. 2007. "Anaxagoras and the Solar Eclipse of 478 BC." *Apeiron* 40: 319–44.

———. 2010. "An Ancient Greek Sighting of Halley's Comet?" *Journal of Cosmology* 9: 2130–36.

Grant, Edward. 2007. *A History of Natural Philosophy: From the Ancient World to the Nineteenth Century.* Cambridge: Cambridge University Press.

Gregory, Andrew. 2000a. "Plato and Aristotle on Eclipses." *Journal for the History of Astronomy* 31: 245–59.

———. 2000b. *Plato's Philosophy of Science.* London: Duckworth.

———. 2007. *Ancient Greek Cosmogony.* London: Duckworth.

Guerlac, Henry. 1963. "Some Historical Assumptions in the History of Science." In *Scientific Change*, edited by A.C. Crombie, 797–812. London: Heinemann.

Guthrie, W.K.C. 1962–1981. *A History of Greek Philosophy.* 6 Vols. Cambridge: Cambridge University Press.

Hahn, Robert. 2001. *Anaximander and the Architects.* Albany: SUNY Press.

Hartner, Willy. 1969. "Eclipse Periods and Thales' Prediction of a Solar Eclipse: Historic Truth and Modern Myth." *Centaurus* 14: 60–71.

Heath, Thomas. 1913. *Aristarchus of Samos: The Greek Copernicus.* Oxford: Clarendon Press.

Heidel, W.A. 1906. "The *Dinē* in Anaximenes and Anaximander." *Classical Philology* 1: 279–82.

———. 1933. *The Heroic Age of Science: The Conceptions, Ideals, and Methods of Science Among the Ancient Greeks.* Baltimore: Williams and Wilkins.

———. 1940. "The Pythagoreans and Greek Mathematics." *American Journal of Philology* 61: 1–33.

Hetherington, Norriss S. 1993. "The Presocratics." In *Cosmology: Historical, Literary, Philosophical, Religious, and Scientific Perspectives*, edited by Norriss S. Hetherington, 53–66. New York: Garland.

———. 1996. "Plato and Eudoxus: Instrumentalists, Realists, or Prisoners of Themata?" *Studies in History and Philosophy of Science* 27: 271–89.
Hölscher, Uvo. 1953. "Anaximander und die Anfänge der Philosophie." *Hermes* 81: 257–77; 358–418.
———. 1965. "Weltzeiten und Lebeskyklus." *Hermes* 93: 7–33.
Huffman, Carl A. 1993. *Philolaus of Croton: Pythagorean and Presocratic*. Cambridge: Cambridge University Press.
———. 2013. "Reason and Myth in Early Pythagorean Cosmology." In *Early Greek Philosophy: The Presocratics and the Emergence of Reason*, 71–98. Washington, D.C.: Catholic University of America Press.
Hunger, Hermann, and David Pingree. 1999. *Astral Sciences in Mesopotamia*. Leiden: Brill.
Hussey, Edward. 1982. "Epistemology and Meaning in Heraclitus." In *Language and Logos*, edited by Malcolm Schofield and Martha Craven Nussbaum, 33–59. Cambridge: Cambridge University Press.
Jacoby, Felix. 1957–1963. *Die Fragmente der griechischen Historiker*. Reprint. 18 Vols. Leiden: Brill.
Kahn, Charles H. 1960. *Anaximander and the Origins of Greek Cosmology*. New York: Columbia University Press.
———. 1970. "On Early Greek Astronomy." *Journal of Hellenic Studies* 90: 99–116.
———. 1979. *The Art and Thought of Heraclitus*. Cambridge: Cambridge University Press.
———. 2001. *Pythagoras and the Pythagoreans*. Indianapolis: Hackett.
Kalfas, Vassilis. 1990. "Criteria Concerning the Birth of a New Science: The Case of Greek Astronomy." In *Greek Studies in the Philosophy and History of Science*, edited by Pantelis Nicolacopoulos. Boston Studies in the Philosophy of Science, vol. 121, 171–85. Dordrecht: Kluwer.
Kamienski, Michael. 1956. "Halley's Comet and Early Chronology." *Journal of the British Astronomical Association* 66: 127–31.
———. 1957. "Researches on the Periodicity of Halley's Comet, Part 3: Revised List of Ancient Perihelion Passages of the Comet." *Acta Astronomica* 7: 111–18.
Keyser, Paul T. 1992. "Xenophanes' Sun (Frr. A32, 33.3, 40 DK6) on Trojan Ida (Lucr. 5.660-5, D.S. 17.7.5-7, Mela 1.94-5)." *Mnemosyne* 45: 299–311.
Kirk, G.S. 1954. *Heraclitus: The Cosmic Fragments*. Cambridge: Cambridge University Press.
Kirk, G.S., J.E. Raven, and M. Schofield. 1983 [1957]. *The Presocratic Philosophers*, 2d edition. Cambridge: Cambridge University Press.
Kingsley, Peter. 1994. "Empedocles' Sun." *Classical Quarterly* 44: 316–24.
Kleingünther, Adolf. 1933. *Prōtos heuretēs: Untersuchungen zur Geschichte einer Fragestellung*. Leipzig: Dieterich. Repr. New York: Arno Press, 1976.
Klowski, Joachim. 1972. "Ist der Aer des Anaximenes als eine Substanz konzipiert?" *Hermes* 100: 131–42.
Kragh, Helge. 1987. *An Introduction to the Historiography of Science*. Cambridge: Cambridge University Press.

———. 2007. *Conceptions of Cosmos: From Myths to the Accelerating Universe: A History of Cosmology*. Oxford: Oxford University Press.

Kuhn, Thomas S. 1957. *The Copernican Revolution*. Cambridge: Harvard University Press.

———. 1996 [1962, 1970]. *The Structure of Scientific Revolutions*, 3d ed. Chicago: University of Chicago Press.

Lakatos, Imre. 1970. "Falsification and the Methodology of Scientific Research Programmes." In *Criticism and the Growth of Knowledge*, edited by I. Lakatos and A.E. Musgrave, 91–196. Cambridge: Cambridge University Press.

Landgraf, Werner. 1984. *On the Motion of Comet Halley*, Max-Plank Institut für Aeronomie no. ESTEC EP/14.7/6184 Final Report. Lindau: Max-Planck Institut für Aeronomie.

———. 1986. "On the Motion of Comet Halley." *Astronomy and Astrophysics* 163: 246–60.

Laks, André. 2006. *Introduction à la "philosophie présocratique."* Paris: Presses Universitaires de France.

———. 2008 [1983]. *Diogène d'Apollonie*, 2d ed. Sankt Augustin, Germany: Academia Verlag.

Laks, André, and Claire Louguet, eds. 2002. *Qu'est-ce que la philosophie présocratique?* Villeneuve d'Ascq: Presses Universitaires du Septentrion.

Laudan, Larry. 1977. *Progress and Its Problems: Toward a Theory of Scientific Growth*. Berkeley: University of California Press.

Lebedev, Andrei V. 1990. "Aristarchus of Samos on Thales' Theory of Eclipses." *Apeiron* 23: 77–85.

———. 1995. "The Cosmos as a Stadium: Agonistic Metaphors in Heraclitus." *Phronesis* 30: 131–50.

Lehoux, Daryn. 2007. *Astronomy, Weather, and Calendars in the Ancient World*. Cambridge: Cambridge University Press.

Leichty, Erle, Maria deJ. Ellis, and Pamela Gerardi, eds. 1988. *A Scientific Humanist: Studies in Memory of Abraham Sachs*. Philadelphia: Publications of the Samuel Noah Kramer Fund.

Leplin, Jarrett. 1984. *Scientific Realism*. 1984. Berkeley: University of California Press.

Lesher, James H. 1992. *Xenophanes of Colophon: Fragments*. Toronto: University of Toronto Press.

Lindberg, David C. 1992. *The Beginnings of Western Science: The European Scientific Tradition in Philosophical, Religious, and Institutional Context, 600 B.C. to A.D. 1450*. Chicago: University of Chicago Press.

Lloyd, Alan B. 1975. Introduction. In *Herodotus, Book II*. Leiden: E.J. Brill.

———. 1976. "Commentary 1–98." In *Herodotus, Book II*. Leiden: E.J. Brill.

Lloyd, G.E.R. 1970. *Early Greek Science: Thales to Aristotle*. New York: W.W. Norton.

———. 1978. "Saving the Appearances." *Classical Quarterly* 28: 202–22. Repr. in Lloyd 1991a, 254–77.

———. 1979. *Magic, Reason and Experience*. Cambridge: Cambridge University Press.

———, ed. 1991a. *Methods and Problems in Greek Science*. Cambridge: Cambridge University Press.

———. 1991b. "The Invention of Nature." In Lloyd 1991a, 417–34.

———. 1991c. "The Debt of Greek Philosophy and Science to the Ancient Near East." In Lloyd 1991a, 281–98.

———. 1992. "Methods and Problems in the History of Ancient Science." *Isis* 83: 564–77.

———. 1996. *Adversaries and Authorities: Investigations into Ancient Greek and Chinese Science*. Cambridge: Cambridge University Press.

———. 2002. *The Ambitions of Curiosity: Understanding the World in Ancient Greece and China*. Cambridge: Cambridge University Press.

Long, A.A. 1974. "Empedocles' Cosmic Cycle in the 'Sixties." In *The Pre-Socratics: A Collection of Critical Essays*, edited by A.P.D. Mourelatos, 397–425. Garden City, NY: Doubleday.

Longrigg, James. 1965. "*Krystalloeidôs*." *Classical Quarterly*, n.s., 15: 249–51.

Maniatis, Yiorgo N. 2009. "Pythagorean Philolaus' Pyrocentric Universe: Its Significance and Contribution to Astronomy and Astrophysics." *ΣΧΟΛΗ: Ancient Philosophy and the Classical Tradition* 3: 401–415.

Mansfeld, Jaap. 1979. "The Chronology of Anaxagoras' Athenian Period and the Date of His Trial I." *Mnemosyne*, 4th ser., 32: 39–69.

———. 1980. "The Chronology of Anaxagoras' Athenian Period and the Date of His Trial II." *Mnemosyne*, 4th ser., 33: 17–95.

———. 1990. *Studies in the Historiography of Greek Philosophy*. Assen, Netherlands: Van Gorcum.

———. 1992a. "*Physikai Doxai* and *Problemata Physica* from Aristotle to Aëtius (and Beyond)." In *Rutgers University Studies in the Classical Humanities*, edited by W.W. Fortenbaugh and D. Gutas, Vol. 5: *Theophrastus: His Psychological, Doxographical and Scientific Writings*, 63–111. New Brunswick.

———. 1992b. *Heresiography in Context: Hippolytus' Elenchos as a Source for Greek Philosophy*. Philosophia Antiqua, vol. 56. Leiden: E.J. Brill.

———. 1999. "Sources." In *The Cambridge Companion to Early Greek Philosophy*, edited by A.A. Long, 22–44. Cambridge: Cambridge University Press.

Mansfeld, Jaap, and David T. Runia. 1997. *Aëtiana: The Method and Intellectual Context of a Doxographer*, Vol. 1: *The Sources*. Philosophia Antiqua 73. Leiden: E.J. Brill.

———. 2009. *Aëtiana: The Method and Intellectual Context of a Doxographer*, Vol. 2: *The Compendium*. 2 parts. Philosophia Antiqua 114. Leiden: E.J. Brill.

Martin, Alain, and Oliver Primavesi. 1999. *L'Empédocle de Strasbourg*. Berlin: Walter de Gruyter.

Marcovich, Miroslav. 1967. *Heraclitus*. Mérida, Venezuela: University of the Andes Press.

———. 1999. *Diogenes Laertius Vitae philosophorum*. 2 vols. Stuttgart: Teubner.

McKirahan, Richard D. 2010 [1994]. *Philosophy Before Socrates*. 2d ed. Indianapolis: Hackett.

BIBLIOGRAPHY

Mejer, Jørgen. 1978. *Diogenes Laertius and His Hellenistic Background*. Hermes Einzelschriften, vol. 40. Wiesbaden: Franz Steiner.

———. 2002. "Eudemus and the History of Science." In Bodnár and Fortenbaugh 2002, 243–61.

Mesturini, Anna Maria. 1987–1988. "L'ecclisi di sole in Empedocle e in Lucrezio." *Studii Italiani di Filologia Classica*, 3rd ser., 5–6: 173–80.

Miller, Richard W. 1987. *Fact and Method*. Princeton: Princeton University Press.

Mittelstrass, Jürgen. 1962. *Die Rettung der Phänomene: Ursprung und Geschichte eines antiken Forschungsprinzips*. Berlin: de Gruyter.

Moran, Jerome. 1973. "Ps-Plutarch's Account of the Heavenly Bodies in Anaximenes." *Mnemosyne*, 26: 9–14.

Morrison, J.S. 1955. "Parmenides and Er." *Journal of Hellenic Studies* 75: 59–68.

———. 1959. "The Shape of the Earth in Plato's *Phaedo*." *Phronesis* 4: 101–19.

Mosshammer, Alden A. 1981. "Thales' Eclipse." *Transactions of the American Philological Association* 111: 145–55.

Mourelatos, Alexander P.D. 1981. "Astronomy and Kinematics in Plato's Project of Rationalist Explanation." *Studies in History and Philosophy of Science* 12: 1–32.

———. 2002a. "La terre et les étoiles dans la cosmologie de Xénophane." In Laks and Louguet 2002, 331–50.

———. 2002b. "Xenophanes' Contribution to the Explanation of the Moon's Light." *Philosophia* 32: 47–58.

———. 2008a [1970]. *The Route of Parmenides*. Rev. ed. Las Vegas: Parmenides Publishing.

———. 2008b. "The Cloud-Astrophysics of Xenophanes and Ionian Material Monism." In Curd and Graham 2008, 134–68.

———. 2012. "'The Light of Day by Night': *nukti phaos*, Said of the Moon in Parmenides B14." In *Presocratics and Plato: Festschrift at Delphi in Honor of Charles Kahn*, edited by Richard Patterson et al., 25–58. Las Vegas: Parmenides Publishing.

Naddaf, Gerard. 2005. *The Greek Concept of Nature*. Albany: State University of New York Press.

Natorp, Paul. 1921. *Platos Ideenlehre: Eine Einführung in den Idealismus*. 2nd ed. Leipzig: Dürr.

Neugebauer, Otto. 1957. *The Exact Sciences in Antiquity*, 2d ed. Providence, RI: Brown University Press.

———. 1972. "On Some Aspects of Early Greek Astronomy." *Proceedings of the American Philosophical Society* 116: 243–51.

———. 1975. *A History of Ancient Mathematical Astronomy*. 3 Vols. Berlin: Springer Verlag.

Newton-Smith, W.H. 1981. *The Rationality of Science*. London: Routledge & Kegan Paul.

Norris, Christopher. 1997. *Against Relativism: Philosophy of Science, Deconstruction and Critical Theory*. Oxford: Blackwell.

North, John. 1995. *The Norton History of Astronomy and Cosmology*. New York: W.W. Norton.

BIBLIOGRAPHY

O'Brien, Denis. 1968. "Derived Light and Eclipses in the Fifth Century." *Journal of Hellenic Studies* 88: 114–27.
———. 1968a. "The Relation of Anaxagoras and Empedocles." *Journal of Hellenic Studies* 88: 93–113.
———. 1969. *Empedocles' Cosmic Cycle*. Cambridge: Cambridge University Press.
O'Grady, Patricia F. 2002. *Thales of Miletus: The Beginnings of Western Science and Philosophy*. Aldershot: Ashgate.
Olmstead, A.T. 1948. *History of the Persian Empire*. Chicago: University of Chicago Press.
Palmer, John. 2009. *Parmenides and Presocratic Philosophy*. Oxford: Oxford University Press.
Panchenko, Dmitri. 1994. "Thales' Prediction of a Solar Eclipse." *Journal of the History of Astronomy* 25: 277–88.
———. 1996. "Thales' Theory of Solar Eclipses and the Birth of Theoretical Science in Early Sixth-Century Ionia." *Hyperboreus* 2: 47–124.
———. 1997. "Anaxagoras' Argument Against the Sphericity of the Earth." *Hyperboreus* 3: 175–78.
———. 1999. "The Shape of the Earth in Archelaus, Democritus and Leucippus." *Hyperboreus* 5: 22–39.
———. 2002. "Eudemus Fr. 145 Wehrli and the Ancient Theories of Lunar Light." In Bodnár and Fortenbaugh 2002, 323–36.
Pannekoek, Anton. 1961 [1951]. *A History of Astronomy*. New York: Interscience Publishers.
Patzer, Andreas. 1986. *Der Sophist Hippias als Philosophiehistoriker*. Freiburg, Germany: Karl Alber.
Pedersen, Olaf. *Early Physics and Astronomy: A Historical Introduction*. 1993 [1974]. Rev. edn. Cambridge: Cambridge University Press.
Pellikann-Engel, Maja E. 1974. *Hesiod and Parmenides: A New View on Their Cosmologies and on Parmenides' Proem*. Amsterdam: Adolf M. Hakkert.
Pendrick, Gerard J. 2002. *Antiphon the Sophist: The Fragments*. Cambridge: Cambridge University Press.
Perilli, Lorenzo. 1992. "La teoria del vortice in Anassimandro e Anassimene: la testimonianza di Epicuro." *Wiener Studien* 105: 5–18.
Pfundstein, James M. 2003. "*Lamprous dunastas*: Aeschylus, Astronomy and the *Agamemnon*." *Classical Journal* 98: 397–410.
Pingree, David. 1992. "Hellenophilia versus the History of Science." *Isis* 83: 554–63.
Pinker, Steven. 1992. *The Blank Slate: The Modern Denial of Human Nature*. New York: Penguin.
Pinto de Oliveira, J. C. 2012. "Kuhn and the Genesis of the 'New Historiography of Science.'" *Studies in History and Philosophy of Science* 43: 115–21.
Podlecki, Anthony J. 1998. *Perikles and His Circle*. London: Routledge.
Popper, Karl R. 1968. *Conjectures and Refutations: The Growth of Scientific Knowledge*. New York: Harper & Row.

BIBLIOGRAPHY

———. 1998. *The World of Parmenides: Essays on the Presocratic Enlightenment*. Edited by Arne F. Petersen and Jørgen Mejer. London: Routledge.

Préaux, Claire. 1973. *La lune dans la pensée grecque*. Brussels: Palais des Académies.

Putnam, Hilary. 1978. *Meaning and the Moral Sciences*. London: Routledge & Kegan Paul.

Ramsey, John T. 2006. *A Descriptive Catalogue of Greco-Roman Comets from 500 BC to AD 400*. Syllecta Classica 17. Iowa City: University of Iowa Press. Repr. with corrections, 2008.

———. 2007. "A Catalogue of Greco-Roman Comets from 500 BC to AD 400." *Journal for the History of Astronomy* 38: 175–97.

Raven, J.E. 1948. *Pythagoreans and Eleatics*. Cambridge: Cambridge University Press.

Reiner, E., and D. Pingree. 1975. *Babylonian Planetary Omens*. Part 1: *Enuma Anu Enlil, Tablet 63*. Bibliotheca Mesopotamica 2.1. Malibu.

Riedweg, Christoph. 2005. *Pythagoras: His Life, Teaching, and Influence*. Translated by Steven Rendall. Ithaca, NY: Cornell University Press.

Rihll, T.E. 1999. *Greek Science*. Oxford: Oxford University Press.

Rochberg, Francesca. 1992. "Introduction: The Cultures of Ancient Science: Some Historical Reflections." *Isis* 83: 547–53.

———. 2004. *The Heavenly Writing: Divination, Horoscopy, and Astronomy in Mesopotamian Culture*. Cambridge: Cambridge University Press.

Rochberg-Halton, Francesca. 1991. "Between Observation and Theory in Babylonian Astronomical Texts." *Journal of Near Eastern Studies* 50: 107–20.

Rosenmeyer, Thomas G. 1956. "*Phaedo* 111C4 ff." *Classical Quarterly*, n.s., 6: 193–97.

———. 1959. "The Shape of the Earth in the *Phaedo*: A Rejoinder." *Phronesis*, 4: 71–72.

Ross, W.D. 1924. *Aristotle's Metaphysics*. 2 Vols. Oxford: Clarendon Press.

Rossetti, Livio. 2011. "Gli onori resi a Talete dalla città di Atene." *Hypnos* 27: 205–21.

Rovelli, Carlo. 2011 [2009]. *The First Scientist: Anaximander and His Legacy*. Translated by Marion Lignana Rosenberg. Yardley, PA: Westholme.

Runia, David T. 1989. "Xenophanes on the Moon: A *doxographicum* in Aëtius." *Phronesis* 34: 245–69.

———. 2008. "The Sources for Presocratic Philosophy." In Curd and Graham 2008, 27–54.

Sambursky, Shmuel. 1956. *The Physical World of the Greeks*. Translated by Merton Dagut. London: Routledge & Kegan Paul.

Santillana, Giorgio de. 1961. *The Origins of Scientific Thought: From Anaximander to Proclus, 600 BC to 300 AD*. Chicago: University of Chicago Press.

Sarton, George. 1927–1948. *An Introduction to the History of Science*. 3 Vols. in 5. Baltimore: Williams and Wilkins.

———. 1952. *A History of Science*. 2 Vols. Cambridge: Harvard University Press.

Schaefer., Bradley E. 2006. "The Origin of the Greek Constellations." *Scientific American* 295:5 (November): 96–101.

Schofield, Malcolm. 1980. *An Essay on Anaxagoras*. Cambridge: Cambridge University Press.

———. 1998. "Anaxagoras." In *Routledge Encyclopedia of Philosophy*, edited by Edward Craig, vol. 1, 249–54. London: Routledge.
Schove, D. Justin. 1948. "Sunspots and Aurorae." *Journal of the British Astronomical Association* 58: 178–90.
Schwabl, Hans. 1966. "Anaximenes und die Gestirne." *Wiener Studien* 79: 33–38.
Sedley, David. 2007. *Creationism and Its Critics in Antiquity*. Berkeley: University of California Press.
Shapin, Steven. 1996. *The Scientific Revolution*. Chicago: University of Chicago Press.
Sider, David. 1973. "Anaxagoras on the Size of the Sun." *Classical Philology* 68: 128–29.
———. 2005 [1981]. *The Fragments of Anaxagoras*. 2d ed. Sankt Augustin, Germany: Academia Verlag.
Skinner, Quentin. 1969. "Meaning and Understanding in the History of Ideas." *History and Theory* 8: 3–53.
Snell, Bruno. 1944. "Die Nachrichten über die Lehren des Thales und die Anfänge der griechischen Philosophie- und Literaturgeschichte." *Philologus* 96: 170–82.
Solmsen, Friedrich. 1965. "Love and Strife in Empedocles' Cosmology." *Phronesis* 10: 109-48.
Stahl, William Harris. 1952. *Macrobius: Commentary on the Dream of Scipio*. New York: Columbia University Press.
Stephenson, F. Richard. 1990. "The Ancient History of Halley's Comet." In *Standing on the Shoulders of Giants: A Longer View of Newton and Halley*, edited by Norman J.W. Thrower, 231–53. Berkeley: University of California Press.
———. 1997. *Historical Eclipses and Earth's Rotation*. Cambridge: Cambridge University Press.
Stephenson, F. Richard, and Louay J. Fatoohi. 1997. "Thales' Prediction of a Solar Eclipse." *Journal for the History of Astronomy* 28: 279–82.
Stokes, Michael C. 1962. "Hesiodic and Milesian Cosmogonies I." *Phronesis* 7: 1–37;
———. 1963. "Hesiodic and Milesian Cosmogonies II." *Phronesis* 8: 1–34.
———. 1965. "On Anaxagoras." *Archiv für Geschichte der Philosophie* 47: 1–19; 217–50.
Tannery, Paul. 1880. "Thalès et ses emprunts à l'Egypt." *Revue Philosophique de la France et de l'Étranger* 9: 299–318.
———. 1893. *Recherches sur l'histoire de l'astronomie ancienne*. Paris: Gauthiers-Villars et fils.
———. 1930 [1887]. *Pour l'histoire de la science hellène*. 2d ed. Edited by A. Dies. Paris: Gauthiers-Villars et Cie.
Taub, Liba. 2003. *Ancient Meteorology*. London: Routledge.
Taylor, A.E. 1917. "On the Date of the Trial of Anaxagoras." *Classical Quarterly* 11: 81–87.
Thurston, Hugh. 1994. *Early Astronomy*. New York: Springer.
Tigner, Steven S. 1974. "Empedocles' Twirled Ladle and the Vortex-Supported Earth." *Isis* 65: 433–47.
———. 1979. "Stars, Unseen Bodies and the Extent of the Earth in Anaxagoras' Cosmogony." In *Arktouros: Hellenic Studies Presented to Bernard M.W. Knox*, edited

by Glen W. Bowersock, Walter Burkert, and Michael C.J. Putnam, 330–35. Berlin: De Gruyter.

Toomer, G.J. 1988. "Hipparchus and Babylonian Astronomy." In Leichty et al. 1988, 353–62.

Toulmin, Stephen. 1967. "The Astrophysics of Berossos the Chaldean." *Isis* 58: 65–76.

Toulmin, Stephen, and June Goodfield. 1961. *The Fabric of the Heavens: The Development of Astronomy and Dynamics*. New York: Harper & Row.

Tucker, Aviezer. 2004. *Our Knowledge of the Past: A Philosophy of Historiography*. Cambridge: Cambridge University Press.

van der Waerden, B. L. 1954. *Science Awakening*. Translated by Arnold Dresden. Groningen: P. Noordhoff.

Vlastos, Gregory. 1953. Review of *Pythagoreans and Eleatics* by J.E. Raven. *Gnomon* 25: 29–35.

———. 1975. *Plato's Universe*. Seattle: University of Washington Press.

von Fritz, Kurt. 1961. "Der Beginn universalwissenschaftlicher Bestrebung und der Primat der Griechen." *Studium Generale* 14: 546–83, 601–36.

———. 1971a. *Grundprobleme der Geschichte der antiken Wissenschaft*. Berlin: Walter de Gruyter.

———. 1971b. *Der Ursprung der Wissenschaft bei den Griechen*. In von Fritz 1971a, 1–334.

von Staden, Heinrich. 1992. "Affinities and Elisions: Helen and Hellenocentrism." *Isis* 83: 578–95.

Wasserstein, A. 1962. "Greek Scientific Thought." *Proceedings of the Cambridge Philological Society*, n.s., 8: 51–63.

West, M.L. 1960. "Anaxagoras and the Meteorite of 467 B.C." *Journal of the British Astronomical Association* 70: 368–69.

———. 1971. *Early Greek Philosophy and the Orient*. Oxford: Clarendon Press.

———. 1980. "The Midnight Planet." *Journal of Hellenic Studies* 100: 206–08.

———. 1994. "*Ab ovo*: Orpheus, Sanchuniathon, and the Origins of the Ionian World Model." *Classical Quarterly* 44: 289–307.

———. 1997. *The East Face of Helicon: West Asiatic Elements in Greek Poetry and Myth*. Oxford: Clarendon Press.

Whewell, William. 1857 [1837]. *History of the Inductive Sciences*. 3d ed. 3 Vols. London: John W. Parker and Son.

White, Stephen A. 2002. "Thales and the Stars." In *Presocratic Philosophy*, edited by Victor Caston and Daniel W. Graham, 3–18. Aldershot: Ashgate.

Wilson, Robert. 1997. *Astronomy Through the Ages: The Story of the Human Attempt to Understand the Universe*. Princeton: Princeton University Press.

Windschuttle, Keith. 2000. *The Killing of History: How Literary Critics and Social Theorists Are Murdering the Past*. San Francisco: Encounter Books.

Wöhrle, Georg. 1993. *Anaximenes aus Milet: Die Fragmente zu seiner Lehre*. Stuttgart: Franz Steiner Verlag.

———. 1995. "Wer entdeckte die Quelle des Mondlichts?" *Hermes* 123: 244–47.

———, ed. 2009. *Die Milesier: Thales*. Berlin: de Gruyter.
———, ed. 2012. *Die Milesier: Anaximander und Anaximenes*. Berlin: de Gruyter.
Woodbury, Leonard. 1981. "Anaxagoras and Athens." *Phoenix* 35: 295–315.
Wright, M.R. 1981. *Empedocles: The Extant Fragments*. New Haven: Yale University Press.
———. 1995. *Cosmology in Antiquity*. London: Routledge.
Yeomans, Donald K., and Tao Kiang. 1981. "The Long-Term Motion of Comet Halley." *Monthly Notices of the Royal Astronomical Society* 197: 633–46.
Zeller, Eduard. 1919–1920. *Die Philosophie der Griechen in ihrer geschichtlichen Entwicklung*. Edited by Wilhelm Nestle, Part 1: *Vorsokratische Philosophie*. 6th ed. Leipzig: O.R. Reisland.
Zhmud, Leonid. 1997. *Wissenschaft, Philosophie und Religion im frühen Pythagoreismus*. Berlin: Akademie Verlag.
———. 2006. *The Origin of the History of Science in Classical Antiquity*. Berlin: de Gruyter.

INDEX OF PASSAGES

Aetius
1.3.1, 49
1.4.3, 171
2.2.1, 64
2.7.1, 106, 135
2.7.7, 94, 193
2.13.1, 49
2.13.2, 188
2.13.4, 72
2.13.5, 190
2.13.8, 76
2.13.9, 191
2.13.10, 64, 65, 66
2.13.11, 188
2.14.3–4, 64, 93
2.14.4, 64
2.15.3, 133
2.15.6, 80, 93
2.15.7, 92, 105
2.16.2–3, 93
2.16.5, 59
2.18.1, 72
2.20.1, 112
2.20.2, 64, 65
2.20.3, 71, 115
2.20.7, 200
2.20.10, 191
2.20.12, 194
2.20.13, 188
2.20.15, 232
2.21.1, 112
2.21.2, 188
2.23, 191
2.23.1, 64, 81
2.23.4, 191
2.24.1, 53
2.24.4, 71, 115
2.24.6, 19
2.24.9, 71, 72
2.25.1, 112
2.25.2, 65
2.25.4, 72
2.25.5, 215
2.25.8, 51
2.25.9, 123
2.25.10, 191
2.25.11, 201
2.27.1, 215
2.28, 207
2.28.1, 72, 112
2.28.4, 232
2.28.5, 18, 51, 89, 195
2.28.6, 112
2.29.3, 207, 232
2.29.4, 19, 195, 196
2.29.5, 72
2.29.6, 55, 207
2.30.1, 194
2.30.3, 200
2.31.1, 152, 187
3.2.2, 165, 199

INDEX OF PASSAGES

Aetius (*continued*)
   3.2.8, 191
   3.2.11, 72, 80
   3.4.4, 70
   3.10.1, 49
   3.12.2, 106
   3.15.7, 106
   4.1.1, 171
   5.19.4, 61

Alcmaeon (DK 24) A4, 93

Anaxagoras (DK 59)
   A1, 126, 140, 145, 159, 165
   A11, 159, 160
   A12, 160, 162, 168
   A17, 141
   A18, 138, 247
   A39, 158
   A42, 123, 138, 145, 171
   A43, 138
   A55, 251
   A76, 132
   A77, 249
   A80, 131, 199
   A81, 165, 199
   A91, 171
   B1, 183
   B6, 183
   B9, 126, 183
   B12, 126, 183–4
   B13, 183
   B15, 184
   B16, 185
   B18, 109, 124, 250

Anaximenes (DK 13)
   A14, 64, 66
   A15, 64
   A16, 64, 65

Anaximander (DK 12)
   A10, 60, 61, 127
   A11, 59, 112, 127
   A18, 59
   A21, 112
   A22, 112
   A30, 61

Antiphon (DK 87)
   B26, 232
   B27, 232
   B28, 232

Apuleius
   Florida 18.32, 55

Aristotle
   On the Heavens
     I.2–3, 221, 234
     II.7, 221
     II.14, 246
     279b12–17, 76
     291b17–21, 113, 220
     291b21–22, 220
     291b22–23, 105, 114
     292a3–6, 220
     292a7–9, 94
     293a20–24, 90
     293b33–294a4, 96, 128
     294a28–33, 48
     295a13–21, 189
     297a8 ff., 246
     297a8–298a20, 95
     297b23–30, 221
   On Coming To Be and Passing Away
     2.7–8, 234
   Metaphysics
     XII.8, 218, 235
     983b25–27, 49
     984a2–3, 48
     984a7–8, 75
     984a18–19, 219
     984b8–11, 219
     989a5–6, 69
     984a11–13, 138
     1073b17–32, 234
     1074b10–13, 219
   Nicomachean Ethics
     VII.1, 181
   Meteorology
     I.3, 221
     I.4–5, 234
     I.6–7, 234
     338a25–b24, 234
     342b27–29, 165, 199
     344b31–34, 160
     345a25–31, 131, 199
     354b33–355a8, 83
     355a13–15, 115
     361b35–362a11, 171
   Physics
     IV, 183
     205a3–4, 76
   Politics
     1329b25–30, 219

INDEX OF PASSAGES

Posterior
  Analytics   87b39–88a2, 219
                 90a15–18, 219
                 93a30–b7, 219
  Fragments   203, 193
pseudo-Aristotle
  Problems   XV.11, 149
                 911b35–912a7, 222
                 912b11–14, 149

Cleomedes   2.4.1–9, 208, 215
                 2.4.12–17, 208, 215
                 2.4.21–32, 215
                 2.4.105, 215
                 2.4.127–37, 215
                 2.5.81–82, 215
                 2.6, 215

Democritus (DK 68)
                 A1, 198
                 A39, 200
                 A86, 133
                 A87, 200
                 A88, 200
                 A89a, 200
                 A90, 200
                 A91, 199
                 A92, 199
                 A96, 106
                 B5, 199

Diogenes of Apollonia (DK 64)
                 A12a, 190, 191
                 A12b, 191, 192
                 A13, 191
                 A14, 191
                 A15, 191

Diogenes Laertius
                 1.22, 47
                 1.23, 47, 48, 50, 51, 56
                 1.24, 51, 55
                 1.27, 49, 50
                 2.1, 59
                 2.7, 140
                 2.8, 145, 194
                 2.9, 165
                 2.10, 159
                 2.12, 126
                 2.23, 233
                 5.48, 89
                 7.145–146, 214
                 7.146, 149
                 8.48, 92
                 8.51–74, 142
                 8.55, 187
                 9.33, 200
                 9.34, 198, 199
                 9.8–11, 73–74
                 9.10, 112
                 9.10–11, 115
                 9.11, 76, 78
                 9.21, 92, 95
                 9.23, 92
                 9.33, 200

Empedocles (DK 31)
                 A7, 138
                 A53, 188
                 A54, 188
                 A60, 188
                 A61, 152, 187
                 A67, 189
                 B14, 187
                 B39, 189
                 B42, 151, 187
                 B43, 91, 186
                 B45, 91, 186, 187
                 B47, 91, 109, 186
                 B48, 188

Epicurus
  On Nature   ΙΑ [33] Arrighetti, 127
  To Herodotus 79–80, 212, 241
  To Pythocles 85–87, 212, 241
                 96, 211

Eudemus
  Fragments   145, 64, 65, 125
                 146, 80, 93, 193

Galen
  On Hippocrates' On the Nature of Man 15.25, 69

Heraclitus (DK 22)
                 A1, 112, 115
                 A12, 112, 115
                 B3, 77

277

INDEX OF PASSAGES

Heraclitus (*continued*)
		Homer	
	B6, 77, 115	*Iliad*	4.75–77, 137, 163
	B12, 73		18.485–89, 57
	B30, 75, 76		18.607–8, 44
	B31, 73, 74		22.318, 92
	B36, 75		23.226, 92
	B60, 73	*Odyssey*	10.508–14, 44
	B76, 75		11.13–15, 44
	B80, 73		13.93–4, 92
	B90, 73		18.219, 100
	B94, 77		
Herodotus	1.74.2, 16	**Ibycus**	fr. 331, 92
	2.19, 170	**Ion of Chios**	
	2.20, 171	(DK 36)	A7, 201
	2.23, 61	**Lucretius**	1.231, 83
	2.25.1–3, 83		1.1090, 41, 83
	4.42, 182		5.621–636, 200
Hesiod			5.660–665, 71, 115
Theogony	22–34, 41		5.720–728, 208
	104–38, 42		5.753–757, 151, 211
	144–45, 124		
	381, 92	**Olympiodorus**	
	671–86, 163	*On the Sacred Art*	24, 69
	713–21, 163	*Meteorology*	67.32–37, 131, 199
	720–28, 44		136.6–11, 78
	728, 59		
	744–54, 80	**Pappus**	
Hippias (DK 86)	A2, 232	*Mathematical*	
	A11, 232	*Collection*	6.555–6, 215
Hippocrates		**Parmenides** (DK 28)	A1, 92, 107, 187
On Breaths	3, 83		A37, 106, 135
On the Nature of Man	1, 69		A40a, 92, 105
Hippolytus			A44, 106
Refutation of All Heresies	1.1.4, 48		B1, 217
	1.1.2–3, 49		B8, 91, 102, 106, 107, 217
	1.6.1–2, 127		B12, 106
	1.6.5, 59, 112		B14, 91, 100
	1.7.4, 64		B15, 91, 104
	1.7.1–8, 63		B49, 106
	1.8.5, 171		
	1.8.6, 126	**Philolaus** (DK 44)	A16, 94, 193
	1.8.7, 184		A19, 193–4,
	1.8.8, 145		A20, 194
	1.8.6–10, 123		B17, 218
	1.8.10, 138	**Philoponus, John**	
	1.14.3, 115	*Categories*	118.4–25, 55
	1.14.1–6, 68–69	*Physics*	262.8–13, 106

INDEX OF PASSAGES

Pindar
  Isthmian Odes   4.24, 92

Plato
  Apology   19b–d, 233
  Cratylus   409a–b, 199, 216
     409b, 132
  Hippias Major   285b–c, 232
  Laws   10.888e–889c, 216
  Phaedo   96a–100a, 233
     99d–e, 149
     100a, 110
     108e ff., 246
     109a, 95
     110b–c, 234
  Protagoras   318d–319a, 232
  Republic   10.616b–617b, 234
  Timaeus   33b–c, 216
     34a–b, 234
     35a–36d, 217
     36b–c, 216
     36b–e, 234
     38b–e, 217
     38c–e, 234
     38d, 217
     39b–c, 217
     39c, 217

Pseudo-Plato
  Epinomis   986e–987d, 217
     987a–d, 94

Pliny the Elder
  Natural History   2.53, 16, 51, 53
     2.149, 160
     18.213, 50
     36.82, 146

Plutarch
  Dinner of the Seven Wise Men   147a, 146
  Lysander   12.1–2, 160, 161–162
     12.4–5, 168
     12.2, 184
  The Face of the Moon   929b, 111, 124
     929c, 111, 200, 215
     932a, 145
     932b–c, 215
  Nicias   23.2–3, 138, 247
  Pericles   32.1, 141
     32.3, 141
  Symposium   730e, 61

pseudo-Plutarch
  Miscellanies   2, 60, 61, 127
     3, 66, 67
     4, 115
     7, 200

Proclus
  On Euclid   352.14–18, 146

Protagoras (DK 80)   A5, 232

Posidonius
  Fragments   122, 215
     123, 215
     124, 215
     125, 215

Ptolemy, Claudius
  Almagest   4.1, 212

Scholia
  On Plato's Republic   498a, 78
  On Basil Peri Geneseōs   92
  On Apollonius Rhodes   A77, 251

Seneca
  Natural Questions   3.14.1, 48
     4a.2.17, 171
     4a.2.22, 171
     7.3.2, 166
     7.5.3, 168

Simplicius
  On the Heavens   294.4–7, 76
     471.2–6, 80, 93, 193
     488.18–24, 234
     493.4 ff., 234
  Physics   23.21–29, 49
     23.33–24.6, 75
     25.19–20, 138
     26.7–10, 139
     149.28–150.4, 75

Stobaeus   1.10.12, 69
     1.22.1d, 94
     1.24.1i, 76
     1.24.2e, 92

Thales (DK 11)   A1, 48
     A5, 16

INDEX OF PASSAGES

Thales (*continued*)
    A14, 48
    A16, 171
    A17a, 49
    A20, 146

Theodoret
  Therapy for
    Greek Diseases 4.23, 65
    4.5, 69

Theon of Smyrna 198.14–18, 50, 53, 64
    199.1–3, 65

Theophrastus
  Fragments
    225B Fortenbaugh, 75
    227B, 187
    227A, 138
    6.4 Wimmer, 236

Thucydides 2.28.1, 212
    4.52.1, 212
    7.50.4, 212

Virgil
  Georgics 2.325–26, 129

Vitruvius 9.2.1, 208
    9.6.2, 208

Xenophanes (DK 21) A33, 115
    A40, 115
    A41, 115
    A41a, 115
    A44, 80
    B28, 70
    B30, 70
    B32, 72
    B34, ?, 179
    B35, 68

Xenophon
  Memorabilia 1.1.13, 15

INDEX

A
Aegean 154, 165, 173, 175
Aegospotami 4, 86, 151, 159–62, 164–5, 175, 191, 231
Aeschylus 172–3, 231, 270
aether 42–3, 82, 105, 123, 126, 161, 183–5, 188, 191, 202, 234, 238
Aetius 18–19, 49, 64–6, 70–2, 80–1, 92–4, 105–6, 112, 115, 187–8, 190–1, 193–7, 199–201, 204, 207–12
air 3, 10, 58, 60, 62–3, 67–71, 76–7, 79, 81–3, 117, 122–3, 160, 183–4, 188, 205
Americas 179–81
Anaxagoras 85, 121–31, 133–42, 145–79, 181–8, 190, 192, 197–202, 204–7, 209–18, 230–5, 237–42, 247–54, 263–5, 271–3
  Athenian period 268
  contributions of 242, 247
  followers of 131–2
  heliophotism of 254
  theory of eclipses 149
  theory of lunar light 216
Anaxagoras and Empedocles 85–7, 107, 125, 142, 144, 152, 174, 176, 253, 270
Anaximander 10, 18–19, 58–62, 71, 79–82, 93, 106, 112–14, 117, 147, 193, 205, 207–9, 265–6, 271
Anaximenes 10, 59, 62–7, 69–71, 74–5, 77, 79, 81–2, 113–14, 117–18, 127, 208, 210–11, 265–6, 273–4
ancient astronomy 80, 218, 253, 264

ancient science 25–6, 28, 37, 255, 268, 271
  new historians of 27, 36
  new historiography of 27, 35
Antiphon 205–7, 209, 211, 213–14, 216, 223, 232, 270
antiphraxis 88, 111, 120–1, 123, 128, 130, 133–4, 156–7, 174–6, 178, 210–12, 214–16, 218–21, 246–51, 253–4
  hypothesis of 145, 158
  theory of 121, 139, 157, 196–7, 211–12, 214, 220, 224, 248, 252
apeiron 70, 183, 261, 263, 265, 267
Apollodorus 140–1
Aristarchus 54–6, 177, 182, 210–12, 215–16, 222–3, 239–40, 262
Aristotelian science 162, 256
Aristotle 48–9, 54, 75–6, 82–3, 94–6, 113–14, 128–9, 131–2, 138–9, 204–7, 211, 216–23, 234–5, 245–6, 252
  account of eclipses 218, 221
  Metaphysics 69, 218, 234–5
  Meteorology 115, 131, 160, 165, 171, 199
asteroids 66, 86, 124, 128, 130, 133, 152, 161, 164, 166, 170, 175, 185, 191–2, 231
astronomers 11, 55–6, 66, 94, 96, 101–2, 104, 149, 158, 211–13, 215, 224, 226, 240, 242
  Greek 53, 146, 250
  Hellenistic 54, 57
  mathematical 215–16
  scientific 13, 223

281

INDEX

astronomical discoveries 92, 102
astronomical theories 28, 47–8, 78, 86, 91,
　96, 100, 109, 118, 124–5, 128, 134, 174–6,
　238–9, 250–1
　early 119, 132
astronomy
　advances of 109, 185
　development of 108, 227, 250
　early 12, 193, 249, 252–3, 272
　early Greek 10, 12–13, 16, 19, 29, 38, 40–1,
　　56, 89, 96, 107–8, 147, 227–8, 247, 269
　events 82, 122, 136, 191, 234
　Greek 8, 14–15, 17, 19, 32, 109, 136, 246, 266
　history of 48, 143, 228, 248, 253
　ideas about 137, 199, 250
　mathematical 19, 89, 236, 240
　new 5, 177, 212, 216
　observational 172, 237
　problems in 176, 251
　scientific 8–9, 14–15, 237
　theoretical 108, 134
Athens 3–4, 9, 140–2, 149, 154–5, 157, 163, 165,
　172–3, 230–1, 233, 244, 249
atmosphere, upper 184–5, 221, 234

B

Babylonian Astronomy 7–9, 17–18, 52–3, 56,
　86–7, 94, 107, 109, 143, 217, 236
Berosus 205–9, 213, 215–16, 224
Bicknell, P. 64–6, 72, 92–3, 168, 261
blocking 47, 53, 55, 85, 88, 120, 124, 128, 130,
　133–4, 145, 151, 174, 185, 219–20
bodies
　astronomical 115, 242
　earthy 65–6, 133, 194, 201, 238
　fiery 63, 167, 188–9
　heavy 59, 117, 122, 126, 128, 130, 163–4, 202, 231
　light 161
　massy 117, 164, 204, 239
　permanent 114, 116, 124, 134, 174, 230, 232,
　　242
　rocky 135, 162
　single 105, 166
　solid 66, 122, 156
　spherical 76, 113–14, 116, 119, 145,
　　220
　stony 67, 126, 135, 162, 184, 202, 231
　sublunary 124

Bodnár, I. 60–2, 112, 261–2, 269–70
Boll, F. 65, 248, 254, 262
Bollack, J. 96, 104, 186, 262
boundless 68, 70, 183, 189
Bowen, A. 14, 48, 53–4, 215, 262
Bowen, A., and Goldstein, B. 9, 14,
　50–1, 236
bowls 10, 74, 76–8, 81–2, 113, 121, 144, 149,
　214, 229
Burkert, W. 44, 90, 92, 93, 166, 207, 262
Burnet, J. 89, 249, 262

C

calendar 7, 9, 14–15, 50, 51, 99, 236, 267
Callias 140–1, 154–5
Clazomenae 4, 149, 151, 154–5, 163,
　165, 176
Cleomedes 56, 208, 215, 224
clouds 8, 10, 61, 63, 67, 69–70, 72–3, 78, 80–1,
　84, 97, 111, 113–14, 144, 156
　glowing 72
Columbus 179–82
comets 4–5, 10, 72, 80, 86, 160, 165–70, 191–2,
　198, 201, 203, 231, 234, 266
condensation 63, 72–5, 81, 115, 210
conflagration 73, 75–6, 135
conjunctions 3, 17, 52, 54, 56, 86, 112, 120–1,
　165–6, 170, 195, 200, 225, 251
constellations 14, 47, 56–7, 64, 229, 237
Cornford, F.M. 89, 234, 263
cosmogony 46, 67, 127
cosmological speculations 15, 232
cosmological theories 5, 45–6, 100, 136, 185,
　229, 253
cosmologists 13, 15, 46, 90, 102, 104, 212
cosmology 15–16, 18–19, 45–7, 90–1,
　103–4, 134–5, 182, 186, 190, 197–8, 200–1,
　216–17, 221–2, 253–4, 265
　fifth-century 201, 203–4
　Greek 45, 90, 266
　scientific 109, 246
cosmos 8, 44–5, 60–2, 67, 79, 126–7, 190, 196,
　216–18, 229, 232–3, 240, 265, 267
counter-earth 193, 195–6, 199, 213–14, 238
Couprie, D. 19, 49, 58–9, 62, 70, 132, 146–8,
　152, 158, 254, 263
Curd, P. 96, 139–40, 263, 269, 271
cycles 51–2, 73, 186, 193

INDEX

D
dark bodies 66, 88, 120, 124, 128, 130, 135, 166, 220, 230–1
Delphi 146–7, 151, 269
Democritus 11, 51, 106, 131, 133, 165–6, 173, 198–201, 223, 231, 263, 270
descriptivists 258–9
Dicks, D.R. 9, 11–14, 50, 57–8, 92–5, 199, 234, 236, 263
Diels, H. 58, 184, 191, 204, 263
Diogenes Laertius 47–51, 55–6, 59, 73, 76, 78, 89, 92, 95, 107, 112, 115, 126, 140, 141, 165, 198–200, 214–15
Diogenes of Apollonia 173, 190–2, 223, 231
Dreyer, J.L.E. 131, 248–9, 264

E
earth
 atmosphere of 66, 133
 circular 44, 80, 171
 disk-shaped 43, 228
 flat 65, 79, 85, 96, 105, 129, 146, 201, 229, 231, 246, 249
 infinite plane of 115, 189
 orbit of 167, 196
 rotation of 98, 272
 shadow of 225–6
 spherical 49, 103, 106, 129, 201, 234
earth diameters 58–9
earthshine 132–3, 207
earthy material 66, 200, 234
earthy natures 63, 66
eclipses
 annular 3, 144, 149–50, 153, 226, 230–1
 crepuscular 130, 198
 explained 123, 125, 137–8, 164
 explanation of 108, 143, 248, 250, 253
 historical 145
 knowledge of 51, 54
 lunar 3, 47, 53, 55, 85, 128–31, 133–4, 184–5, 195–6, 198–9, 207–8, 210–14, 219–20, 225–6, 250–1
 maximum 144, 152, 156
 nature of 138, 225
 solar 3, 16–18, 47, 51–6, 85–6, 119–21, 143–4, 152, 156–7, 159–60, 174–5, 209–10, 220–1, 225–6, 261–2
 total 131, 144, 149, 153, 159, 226, 230

Egypt 18, 49, 58, 146, 170, 172–4, 182, 229, 272
Egyptians 18, 50–1, 56–7, 86, 91, 94
Empedocles 10, 64, 85, 89, 91, 136–9, 142, 151–2, 157–8, 186–90, 194–5, 205–7, 209, 248, 253
empirical confirmation 175–6, 178, 245
Epicurus 127, 211–12, 241
equinoxes 13, 47, 50, 263
Eudemus 48, 50, 53–4, 64–5, 92, 125, 193, 262, 269
Eudoxus 9–10, 13–15, 218, 223, 235, 237, 251, 266
evaporations 63, 67, 73, 210, 229
evening star 92–4, 105, 230, 252
explanations, astronomical 185, 212

F
fifth century BC 19, 54–5, 86–8, 90, 92, 94, 109, 135–6, 143, 175–6, 186, 191–2, 201, 203–4, 240
fixed stars 14, 80, 93, 166, 188–9, 192–3, 210, 229
floods 170–3
Fortenbaugh, W. 138, 262, 268–70
Furley, D. 46, 217, 264

G
geometry 18, 110, 119–20, 214, 225, 227, 239
Gershenson, D., and Greenberg, D. 124, 251, 264
gnomon 50, 146, 273
Goldstein, B. 14, 264
Graham, D. 9, 19–20, 46, 55, 67, 69, 144, 149, 167–70, 173, 186, 233, 244, 253, 265
Gregory, A. 46, 127, 221, 234, 265
Greece 4, 18, 27, 39, 95, 99, 146, 151, 157, 168, 170–1, 236–7, 253
Guthrie, W.K.C. 54, 71, 73, 91, 96, 140, 187, 251, 265

H
Hahn, R. 19, 58–9, 265
Halley's Comet 166–9, 231, 267
Hartner, W. 52–3, 265
hearth 192–3
Heath, T. 56, 64, 66, 91, 93, 177, 211, 234–5, 239, 249, 254, 265

INDEX

Heidel, W.A. 90, 127, 249, 265
heliophotism 87–9, 96–7, 104–5, 109–12, 116–25, 134, 156–7, 174–6, 185–7, 197–8, 205–9, 211–15, 220–4, 246–50, 253–4
   acceptance of 197, 206, 234, 251
   application of 120, 222
   hypothesis of 122, 230
   implications of 120, 122–4, 128, 134, 136, 159
   recognition of 108, 124
hemisphere 76–7, 113, 188, 226
Heraclitus 10, 51, 73–9, 81–2, 103, 112–15, 144, 156, 205, 207–9, 211, 213, 216, 229, 266
Heraclitus on Solar Eclipses 262
Herodotus 16, 51, 53, 83, 141, 153–4, 170–1, 173, 182, 212, 241, 267
Hesiod 44–5, 48, 50, 59, 92, 105, 163, 171, 270
Hintz, E. 144, 149, 167–8, 170
Hipparchus 13, 215, 223, 235, 237
Hippolytus 49, 62, 65, 67–9, 71, 115, 124–6, 138–9, 183
historians of philosophy 259–60
historians of science 9, 12, 89, 257–8, 260
historiography 25, 33, 37, 39, 273
   philosophy of 25
historiography of science 247, 266
history of science 25, 28, 32, 35, 39, 249, 265–6, 269–71
Huffman, C. 90, 94, 194–6, 266

I

idiophotism 205–7, 209
illumination 97, 104, 123, 125, 132–3, 138, 189, 203
instrumentalists 21, 23, 31, 266
interposition 88, 120, 130, 196, 219, 225, 240, 250, 252

K

Kahn, C. 12, 19, 53, 58, 75, 76, 90, 96, 139, 266
Kamienski, M. 167, 266
Kirk, G.S. 47, 54, 65, 69, 71, 75–6, 78, 90, 145, 266
Kragh, H. 25, 34, 36, 266
Kuhn, T. 20, 27, 30, 220, 234–5, 258–9, 267, 270

L

Laks, A. 35, 192, 265, 267, 269
Landgraf, W. 167, 267
Lebedev, A. 54, 77, 267
Lesher, J. 69, 70, 267
light
   borrowed 85, 91
   earth's blocking of 197
   moon's source of 88, 108
   natural 131–2, 207
   reflected 131, 221, 224, 252
   residual 131–2
   secondary 132–3, 184, 207, 214–15
   source of 64, 82, 91, 120, 185, 192, 202, 221, 239
Lithic Model 134–5, 164, 175, 192, 200–1, 204, 232, 235
Lloyd, A.B. 171, 267
Lloyd, G.E.R. 8–9, 11, 17, 26, 32, 36–7, 45, 53, 56, 94, 244, 256, 259, 267–8
lunar eclipses. *See* eclipses, lunar
lunar light 57, 91, 130, 197, 200, 209, 212, 216, 238–9, 254, 270
   source of 51, 55, 130
lunar month 51, 98–9, 110, 116, 119–21, 194, 225

M

Macrobius 56, 272
Mansfeld, J. 48, 125, 139–42, 268
Mansfeld, J., and Runia, D. 19, 64, 76, 89, 93, 191–2, 195–6, 201, 204–5, 207, 210, 213
McKirahan, R. 59, 268
Mejer, J. 48, 59, 269
meteorites 66–7, 80, 159, 161, 163–4, 202, 231, 248, 273
meteoroids 66, 201
Meteorological Model 78, 84, 122, 135, 201, 232
meteors 5, 66, 80, 86, 159–60, 162–6, 170, 173, 175–6, 191, 202, 231, 234
Meton 9, 14, 236–7
Metrodorus 89, 195, 206, 209
Milky Way 131–3, 198
model
   mathematical 8, 14, 55, 178, 234–7
   new 109, 135–6

INDEX

moon
  crescent 51, 112, 132–3
  full 98–9, 112–13, 118, 121, 214, 219–20,
    225–6, 230
  half 77, 99, 112–13, 220
  light of 19, 87, 89, 100, 102–3, 108–9,
    119, 124, 178, 187, 204, 207–8, 220, 243
    245
  new 51, 54, 98, 111–12, 116, 121, 123, 132, 145,
    153, 156, 159, 175, 213, 230
  phases of 108, 110, 243
  position of 108, 110
  shadow of 148, 152
  shape of 99, 114, 116
  surface of 132, 207
morning star 92, 94, 105, 230, 252
Mosshammer, A. 52–3, 269
motion 18, 49, 55, 62, 79, 82, 101–2, 105, 117,
  135, 180–1, 183–5, 216–18, 221, 235
Mourelatos, A.P.D. 32, 70, 72, 91, 97, 115, 127,
  145, 268–9

N
natural philosophers 5, 13, 78, 87, 138, 178, 217,
  229, 248
natural philosophy 11, 49, 68, 89–90, 170, 182,
  191, 216, 232–3, 253, 265
Neugebauer, O. 13, 49, 53, 250, 269
New World 179–81
Nile 170–1, 173, 265
  floods of 170, 231
north pole 47, 56–7, 59, 169, 229
nucleus 66, 71–2, 115

O
objective knowledge 25, 37–8, 40
objectivity 6, 23–4, 26, 28, 32–3, 37–8, 257
O'Brien, D. 131, 139, 186, 270
observations, astronomical 119, 143, 234
occultation 133, 156, 220, 225–6
O'Grady, P. 17, 47, 49–50, 52–3, 56, 270
Old World 179–80
ontology 103, 107, 182
orbits 47, 55, 58–9, 71, 77, 86, 112, 116–17,
  124, 135, 146–7, 166, 216–18, 230–1,
  238–9
  moon's 112–13, 196

P
Panchenko, D. 17–18, 64, 96, 125, 129, 131,
  149, 270
parapēgma 14, 50
Parmenides 89–92, 94–7, 99–111, 114,
  123–4, 134–5, 137, 174–7, 187, 204–7, 209,
  216–17, 246–9, 252–3, 262–3
Peloponnesus 123, 145–9, 151–2, 154–5, 157–8,
  164, 230–1
Pericles 138, 141, 173, 244
phenomena, astronomical 86–7, 241, 243
Philolaus 12, 90, 94–5, 109, 192–9, 206,
  210–12, 214, 216, 223, 231, 238
philosophy, Greek 89, 251, 265, 268
philosophy of science 23, 25, 28, 252
Phoenicians 56–7, 229
Pingree, D. 27–8, 36, 52, 94, 258–9, 270–1
Piraeus 154–6, 230
plane 62, 117, 167, 222, 235
  infinite 70–1, 79, 118
planetary motion, mathematical models of
  239–40, 246
planets 7, 13, 80, 92–4, 133, 149, 165–6, 188,
  192–3, 201, 203, 216–18, 231, 238, 248
  conjunction of 165–6
  five visible 80, 94, 166, 193, 216–17
  motion of 14–15, 240
Plato 9–10, 14–15, 24, 32–3, 38, 65, 110, 132, 149,
  199, 213–14, 216–17, 233–4, 246, 263–4
  Timaeus 216–17, 234
Pliny the Elder 160–1, 165–7
  Natural History 16, 50–1, 53, 146, 160
Plutarch 107, 111, 124, 138–9, 145, 161–3, 167,
  170, 184, 200, 215, 242, 247, 251, 262
Popper, K. 70, 92, 96, 100–4, 107, 121, 270
Posidonius 214–15, 264
practices, scientific 46, 257, 260
Presocratic Philosophers 9–13, 15–17, 39–40,
  94, 96, 111, 192, 221, 223–4, 238, 243,
  248–9, 251–2, 261, 266
Presocratic Philosophy 10, 17, 248, 271, 273
Ptolemy 8, 177, 212, 215, 223, 235
Pythagoras 18–19, 29, 39, 87, 89–90, 107, 109,
  195, 205–8, 249–50, 254, 266, 271
Pythagoreans 15, 18, 89–90, 92, 193–5, 207,
  209–10, 213–14, 249–50, 266, 271, 273
  younger 195–7, 213

INDEX

Q
quenching 72, 210, 213

R
Ramsey, J. 167, 169, 271
rarefaction 73–5, 81
Raven, J.E. 47, 54, 65, 69, 71, 89–90, 266, 271, 273
revolution 39, 122–3, 126, 161, 183–4, 189, 193, 258–9
rings 58–61, 78–9, 81–2, 114, 117, 127, 229
rivers 70, 73, 78, 170
Rochberg, F. 26, 28, 36, 37, 52, 91, 257, 271

S
Sambursky, S. 46, 52, 250, 271
Sarton, G. 249–50, 271
science
  development of 46, 244
  discoveries of 9, 18, 114, 181, 240
  early Greek 8, 28, 255, 267
  empirical 6, 110
  goal of 21
  Greek 44, 244, 252–3, 258, 264, 271
  historian of 26
  history of 22–8, 32, 34, 36–7, 240, 245, 255–60
  modern 16, 28, 110, 255, 258, 260
  new historiography of 27, 39
  normal 19–20, 258–9
  old-time history of 29, 38
  real 27, 178, 220
  western 9, 19, 27, 243, 246, 253, 261, 267
scientific change 20, 265
scientific explanation 19, 244–5
scientific knowledge 18, 24, 32, 37, 40, 219, 243, 249, 270
scientific method 21, 26–7, 257–9, 264
scientific philosophy 19, 265
scientific progress 9, 20, 28–9, 31, 38–40, 237, 244
scientific realism 21–2, 32, 262, 267
scientific research 20–1, 40, 212, 267
Scientific Revolution 39, 272
scientific theories 21–2, 31, 96, 111, 135, 233, 238, 241, 245
shadows 65, 101, 103, 105, 112–13, 130, 146, 148–9, 151, 156–7, 187–8, 213–14, 219–20, 225, 230–1
Sider, D. 127, 139–40, 147, 158, 272
Simplicius 49, 75, 76, 93, 107, 138–9, 193, 217, 234
snows, melting of 170–1
Socrates 1, 10, 15, 140, 233, 244, 265, 268
solar eclipses. *See* eclipses, solar
solar light, reflected 133, 207
solstices 47, 50, 64, 83, 123, 191–2, 236, 263
sophists 232–3, 270
Sparta 146–7, 151
sphere 59, 107, 129, 145, 184, 186, 189, 193, 208, 216–18, 235
sphericity 87, 95–6, 103, 105–7, 217, 221–2, 246, 249, 270
spheroid 174, 185, 190, 202–3
stars 7, 47–51, 59–61, 63–6, 72, 79–81, 83–4, 105, 123–4, 127, 130–3, 184–5, 188–9, 191–2, 200–1
  observations of 7, 237
  shooting 4, 10, 72, 80, 84, 167
  wandering 80, 95, 165, 188, 193
Stephenson, F. 17, 46, 143, 167, 272
Stoics 205–6, 208–9, 213–16, 223
Stokes, M. 45, 83, 162, 272
stones 10, 63, 70, 81, 126, 159–63, 167, 185, 190–1
summer 50, 74, 155, 167, 170–1
sun 54–68, 70–2, 76–81, 83–9, 97–100, 108–12, 114–21, 127–34, 144–52, 156–60, 184–8, 190–4, 196–202, 208–11, 224–32
  diameter of 146–7
  light of 47, 85, 88, 110–11, 119–21, 124, 128, 130–4, 156, 185, 198–9, 207, 219–20, 222, 230
  rays of 133, 151, 174, 186–8, 196, 202, 215
  rising 4, 98–9
  setting 97–9, 105, 129
sunset 98, 129, 198–9
surface, earth's 53, 117, 132, 144, 159, 188, 202, 231
syllogism 219–20
symmetry, rotational 60–1, 106, 217–18

INDEX

T

Tartarus 43–5, 80, 228
Thales 16–19, 29, 46–59, 79, 86–7, 89, 93–4, 109, 117–18, 146–7, 170–1, 205–6, 208–11, 248, 262–4, 270
   prediction of 269–70, 272
Theophrastus 75–6, 88–9, 93, 125, 138–9, 158, 187, 204, 236, 268
Tigner, S. 128, 185, 272
Tucker, A. 25, 33, 35, 273

U

umbra 130–1, 144, 148–9, 152, 158, 230
Ursa Minor 56–7, 229

V

vapors 47, 67, 70, 74, 83, 115, 118, 135, 144, 229
Venus 80, 93–4, 96, 133, 168
Vlastos, G. 9, 11, 32, 90, 217, 234, 253, 273
von Fritz, K. 251, 252, 273
von Staden, H. 26, 28, 255, 257, 273
vortex 126–8, 133, 135, 183, 185, 189–90, 202–3, 235

W

water 10, 45, 47–9, 59, 62–3, 69–70, 73–5, 81, 149, 171, 189–90, 206, 214, 234
water-clock 56
West, M.L . 44, 92, 93, 94, 147, 167, 273
winds 61–4, 66–7, 70, 74, 81–2, 84, 122, 126, 160, 171–2, 228–9
Wöhrle, G. 48, 55, 58, 64–7, 92, 273
world-order 60, 69, 73, 75–6
Wright, M.R. 46, 142, 188, 274

X

Xenophanes 10, 51, 68–72, 76, 79–82, 88, 97, 103, 105, 111, 113–14, 116–18, 205, 207–8, 210–11

Y

Yeomans, D., and Kiang, T. 167, 274

Z

Zeller, E. 127, 139, 274
Zhmud, L. 23, 54, 90, 274